Seth Godin is the bestselling author of *The Dip*, [illegible] *Permission Marketing* and *Small is the New Big*, among other books, and is one of the most popular business bloggers in the world. He holds an MBA from Stanford University and has been called 'the Ultimate Entrepreneur for the Information Age' by *Business Week* magazine.

IF YOU'VE ENJOYED *MEATBALL SUNDAE*,
CHECK OUT THESE OTHER THOUGHT-PROVOKING,
BESTSELLING BOOKS BY SETH GODIN:

Tribes

The Dip

Small Is the New Big

All Marketers Are Liars

Free Prize Inside!

Purple Cow

Unleashing the Ideavirus

Permission Marketing

Survival Is Not Enough

The Big Red Fez

AND CHECK OUT THESE FREE E-BOOKS (GOOGLE 'EM):

Knock Knock

Who's There

Everyone's an Expert

The Bootstrapper's Bible

meatball sundae

HOW NEW MARKETING IS TRANSFORMING THE BUSINESS WORLD (AND HOW TO THRIVE IN IT)

SETH GODIN

AUTHOR OF THE DIP AND PURPLE COW

PIATKUS

PIATKUS

First published in Great Britain in 2008 by Piatkus Books
This paperback edition published in 2009 by Piatkus Books
First published in the US in 2007 by Portfolio, a member of
Penguin Group (USA) Inc.

A CIP catalogue record for this book
is available from the British Library

ISBN 978-0-7499-2948-0

Designed by Daniel Lagin

Data Manipulation by Phoenix Photosetting, Chatham, Kent
www.phoenixphotosetting.co.uk

Printed and bound in Great Britain by
CPI Mackays, Chatham, ME5 8TD

Papers used by Piatkus are natural, renewable and recyclable products made from wood grown in sustainable forests and certified in accordance with the rules of the Forest Stewardship Council

Piatkus Books
An imprint of
Little, Brown Book Group
100 Victoria Embankment
London EC4Y 0DY

An Hachette Livre UK Company
www.hachettelivre.co.uk

www.piatkus.co.uk

A Sundae?

Maybe this is familiar. It is to me, anyway:

You go to a marketing meeting. There's a presentation from the new Internet-marketing guy. He's brought a fancy (and expensive) blogging consultant with him. She starts talking about how blogs and the "Web 2.0 social media infrastructure" are just waiting for your company to dive in. "Try this stuff," she seems to be saying, "and the rest of your competitive/structural/profit issues will disappear."

In the last ten years, the Internet and radical changes in media have provided marketers everywhere with a toolbox that allows them to capture attention with seemingly little effort, planning, or cash. Six years after the dot-com boom, there are more Web sites, more e-mail users, and more viral ideas, online and offline, than ever before. There are hundreds of cable TV networks and thousands of online radio stations. Not to mention street marketing, e-mail marketing, and MySpace.

Corporations, political parties, nonprofits, job seekers, and yes, even people looking for love, are all scrambling around, trying to exploit the power of these new tools. People treat the New Marketing like a kid with a twenty-dollar bill at an ice cream parlor. They keep wanting to add more stuff—more candy bits and sprinkles and cream and cherries. The dream is simple: "If we can just add enough of [today's hot topping], everything will take care of itself."

Most of the time, despite all the hype, organizations fail when they try to use this scattershot approach. They fail to get buzz or traffic or noise or sales. Organizations don't fail because the Web and the New Marketing don't work. They fail because the Web and the New Marketing work only when applied to the right organization. New Media makes a promise to the consumer. If the organization is unable to keep that promise, then it fails.

New Marketing—whipped cream and a cherry on top—isn't magical. What's magical is what happens when an organization uses the New Marketing to become something it never used to be—it's not just the marketing that's transformed, but the entire organization. Just as technology propelled certain organizations through the Industrial Revolution, this new kind of marketing is driving the right organizations through the digital revolution.

You can *become* the right organization. You can align your organization from the bottom up to sync with New Marketing, and you can transform your organization into one that thrives on the new rules.

For Natalie

A Meatball Sundae

Messy, disgusting, ineffective. The result of combining two perfectly good items that don't go well together.

The meatballs are the basic staples, the things that people need, the stuff that used to be marketed quite effectively with TV ads and other mass-market techniques.

The topping is the New Marketing. MySpace, Web sites, YouTube, permission marketing, and viral techniques are all part of the magic that makes up the top of the sundae.

It's not an accident that almost all the brands, products, and careers that have succeeded with New Marketing are brand-new and fresh. The New Marketing demands more than a meatball. It insists on a reinvention of the entire organization and the products it creates. Marketing is now about a lot more than just the yodeling. It's about the entire package. What you say as much as how you say it. New Marketing is our future. Unfortunately, it doesn't work so well with meatballs.

1964

My best advice: Make average stuff for average people. Get retail shelf space. Use every penny you've got left to buy TV time. Sell as much as you can. Repeat with a different product.

2010

My best advice: Realize that most businesses are still living in 1964. Time to get in sync with the New Marketing.

Executive Summary

What business are you in?

Marketing determines the answer to that, and marketing just changed.

1. For a century, successful organizations have been built around traditional marketing tactics. Marketing was an expensive investment, but it was worth it. Take an average product for average people and spend enough money promoting it, and you were likely to make a significant profit. TV was a miracle worker, and the secrets to leveraging the miracle were an efficient factory, consistent products, and the willingness to spend money to tell people about those products.
2. The New Marketing has really shaken up traditional marketers. First, excessive clutter and dozens of new media alternatives have ended the guaranteed effectiveness

of television. Just as the old rules were changing, new techniques—ranging from Web sites to Google ads to digital word of mouth—came along and picked up the slack. To a traditional marketer, an effective New Marketing technique is an even better miracle than TV. Very fast results at almost no cost!

3. But the New Marketing doesn't work for everyone, and it doesn't work as well as some would like. The question marketing consultants are asked all the time is, "But how can we make this new stuff work for *us*?" And there's frustration because this is exactly the wrong question.

This is a book about the *right* question. Not "How do we use the cool new tools to support our existing structure" but "How do we become an organization that thrives because of the New Marketing?"

The wrong question is, "How do we use this New Marketing to maintain business as usual?" The right question is, "If we're not growing the way we'd like to, how can our business itself be altered?"

Ask not what the New Marketing can do for you.
Ask what you can do to thrive with the New Marketing.

Successful organizations got that way by leveraging what worked. They built factories and gave people (and the media) what they wanted, in quantity. Average stuff for average people, delivered straight from the store to you. I call this a *meatball*, a commodity, a branded item of little differentiation and decent quality. We've always needed meatballs. Call them staples or commodities or the basic building blocks of civilized society, but we need them advertised, and we need them in quantity.

Yes, you can still sell plenty of meatballs, but if you're in the meatball business, bad news looms. The media you depended on to sell average stuff to average people is fading. Network TV, newspapers, telemarketing, and cold-calling are all in trouble. Tactics from the New Marketing are taking the juice out of the Old Marketing, without completely replacing it. Traditional organizations can't thrive if they rely on the New Marketing to do their Old Marketing for them.

So, we're at a crossroads. Down one path is ever-increasing frustration as harried marketing VPs are challenged to use some of that New Marketing "Internet stuff" to prop up traditional organizations. Down the other path are nimble, intelligent organizations that are poised and prepared, ready to be propelled by the fresh tactics of the New Marketing.

The New Marketing is lousy at selling meatballs. Today, growth comes from an integrated approach, one that combines the New Marketing tactics with fundamentally different products and services.

Are you stuck with meatballs? Everyone knows how to make and market a meatball. That approach appears safe and straightforward, but it doesn't lead to growth anymore. You can't grow with meatballs because they're ubiquitous. The New Marketing is whipped cream and a cherry, a collection of techniques that offer huge payoffs—but the New Marketing works only for organizations that can get in alignment, stop making meatballs, and start making something that goes very well indeed with hot fudge and marshmallow sauce.

> Fourteen trends are completely remaking what it means to be a marketer. And while these trends are transforming organizations that have the right products and the right approaches, they are crippling the organizations that are stuck with nothing but meatballs. Once again, marketing is transforming what we make and how we make it.

Shortcuts

What am I, stupid?

You would think I'd learn a lesson. My short books sell much better than my long ones. So why not make this book really short? Two reasons. First, because there's a lot of juicy stuff here, tactics you can use right now, stories that can inspire change. Second, because I'm asking a lot out of you once you're finished reading.

Business as usual isn't working anymore.

If I'm going to ask you to radically change the strategy you're using to support your marketing, I need to lay out enough information to make it easy for you to sell yourself (and your peers) on taking action.

The Internet has taught us all to read paragraphs, not chapters, to look for quick hits of insight and then to surf on. This book is formatted to support that approach to acquiring information.

You can read this book in ten minutes. Or you can take two hours or even a few days. Up to you. If you want to know what it's about, just read the executive summary. If you want the high points, jump from one boxed item to another. Or, if you want to dive in deep, grab a pen, scribble in the margins, and start right here.

Contents

Part 1

THINKING ABOUT THE MEATBALL SUNDAE

Before, During, and After

Before Advertising, there were hundreds of thousands of companies. And all of them looked the same. They were small and local, and they built things by hand. Most of these companies failed to make the transition to the next era. They underinvested in marketing and weren't willing to shift gears from bespoke to mass.

This was the blacksmith or the local general store. This was a traveling musician or the local physician. Sure, there are a few left, but it's not a particularly sexy way to make a living.

During Advertising, companies looked the way most of them look today. They made average products for average people, advertised heavily, and created in bulk.

Imagine Tesco or Marks & Spencer. Consider the Red Cross or your local health insurance giant. All are large organizations designed to work well with the masses. This is the sort of organization that investors invest in, that CEOs yearn to run.

Why are we so quick to imagine that most of the companies that work so well with mass (mass marketing, mass markets—the DA era) will seamlessly transition to the After Advertising (AA) regime?

After Advertising, organizations will look as much like the DA companies as the DA companies resembled those from the BA era. In other words, not very much.

It's human nature to imagine that the future will be just like the present, but with cooler uniforms and flying cars. That's not what's happening. Instead, the landscape of tomorrow (and the day after tomorrow) is fundamentally changed from the environment that drove commerce and organizations for the last hundred years.

The organization you work for today was almost certainly invented and optimized and evolved to work perfectly in a world of retailers, media ad buys, local factories, and commodity products.

When those things are altered forever, what will happen to you?

The Foundations of the New Marketing

The New Marketing isn't a single event or Web site or technology. It's based on a combination of more than a dozen trends, each of which is changing the way ideas are perceived and spread. Here are the trends I'll be referring to throughout the book:

- Direct communication and commerce between producers and consumers
- Amplification of the voice of the consumer and independent authorities
- The need for an authentic story as the number of sources increases
- Extremely short attention spans due to clutter
- The Long Tail
- Outsourcing
- Google and the dicing of everything
- Infinite channels of communication

- Direct communication and commerce between consumers and consumers
- The shifts in scarcity and abundance
- The triumph of big ideas
- The shift from "how many" to "who"
- The wealthy are like us
- New gatekeepers, no gatekeepers

I define *Old Marketing* as the act of interrupting masses of people with ads about average products. Masses of people could be processed quickly and cheaply, and some would respond to your message and become customers. The key drivers of this approach were a scarcity of choice and a large resource of cheap attention.

New Marketing leverages scarce attention and creates interactions among communities with similar interests. New Marketing treats every interaction, product, service, and side effect as a form of media. Marketers do this by telling stories, creating remarkable products, and gaining permission to deliver messages directly to interested people.

New Marketing treats every interaction, product, service, and side effect as a form of media.

We Make What We Can Sell

It's always been this way: Organizations match the medium.

In 1984, I went to Toy Fair for the first time. Toy Fair is an annual convention for the toy industry, and in those days (back when there were actual toy stores), every major toy company and every major toy buyer attended. In secret meetings with toy buyers, the toy companies would show off the lineup for the coming year. Here's the thing: Toy Fair was in February, *nine months* before the Christmas shopping season. Why such a long lead time? Because the toy companies didn't show off the toys. They showed off the commercials. The commercials that got the best reception led to toys getting made; the rest were canceled.

In the U.S.A., if you buy a pair of chinos from L. L. Bean, drag them behind your car for a few hundred miles, pour grape juice on them, and then cut them with scissors, you can send them back to the store for a full refund. Why on earth would L. L. Bean have a policy that allows such abuse?

Even worse: Hundreds of people a year visit a Sears store, buy a roomful of furniture, throw a party, and return the beer-stained furniture the next day for a full refund. Sears doesn't like it, but it's part of doing business. Why?

In 1912, L. L. Bean started his mail-order business by mailing a circular to men in Maine who had hunting licenses. What Bean discovered early on was this: While mail order saved him money on rent, people were hesitant to buy something unknown and untouched. As a result, Bean put a substantial portion of the money he saved on rent into customer service and his guarantee. He realized that this investment enabled his business to grow.

Sears understood the same thing. As they rolled out their dry goods offerings via catalog, they discovered that they could sell just about anything (even houses) if they offered a guarantee that people believed in.

The lesson isn't that every mail-order business needs a guarantee. The lesson is that specific marketing models require specific organizational models to back them up. Which comes first? You find an organizational model that will take advantage of the marketing tools available to you.

Microsoft built its monopoly on two pillars. The first was a relationship with the companies that made the computers. The hardware companies decided which operating systems were installed on their machines. Courting IBM—and then

Dell and others—ensured Microsoft's dominance. As long as the people who made the computers chose to install Microsoft's operating system, Windows would dominate and Microsoft would win.

The second pillar was an army of tens of thousands of independent consultants, VARs, and professionals whom Microsoft trained and catered to. Smart management realized that all Microsoft had to do for twenty years was to support one pillar or the other. When both pillars were fully supported, Microsoft won.

Microsoft's focus on computer companies and VARs is marketing, the same way an L. L. Bean catalog is marketing, the same way a TV commercial is marketing.

Business growth doesn't come from your factory; instead, it comes from satisfying the people who can best leverage your ideas. From Microsoft to Bullwinkle, there are centuries of data that confirm one big idea: We make what we can sell.

What do marketers do?

We spread ideas.

We tell stories people want to hear and believe.

We translate emotion into action.

We close the sale.

We make things people want to buy.

We use the best available medium to reach the right people at the right time.

Marketers run things. They always have. Sometimes, though, the people running things don't realize that they are marketers.

When Geico changed the structure of its life insurance to make it easier to sell by phone, they were making a marketing decision. The change involved the entire structure of the organization, but it was done to take advantage of a particular market opportunity. Geico's new structure easily supports an annual ad budget of half a billion dollars at the same time that they offer the lowest prices in most markets. Geico doesn't gain market share because they spend a lot on advertising. They gain market share because the marketing is integral to their organization.

When Microsoft engineers choose the chipset for a new operating system or make critical decisions about the compatibility of an upgrade to a mail server, they are making a marketing decision.

And when Quaker Oats tested a commercial before they made the cereal, they were making a marketing decision as well.

Marketing Tools That Work

Entire industries have been built around effective marketing tools. TV ads, for example, require factories to make items in sufficient quantity to pay for the commercials. They require shelf space, so that when hordes of people go to the store to buy the advertised item, it's there. And they require average products for average people because TV reaches everyone.

Compare this to direct mail. Direct mail permits specialty retailers to thrive in a way that they never could if they were stuck in a strip mall. Direct mail hates products that appeal to everyone. Direct mail can't afford to go to everyone. It needs to be sent directly to just the right people, on just the right day. Direct mail requires teams of people to answer the phone in the middle of the night. It requires a sales force far more trained than a retailer's. And it requires low inventory, because it's too expensive to keep the warehouse full of an item that's not selling.

For a long time, the interruption tools available to marketers were fairly limited. You could interrupt people with advertising (in newspapers, radio, television, etc.). You could hire a sales force. You could develop packaging that stood out. Not only were your choices limited, but most of the tools had been around for a generation or more. Your business was based on running TV ads or handing DJs' payola. Your organization had a sales force; it always had and it always would. Or you didn't have a sales force, because you relied on a rep firm. Your chain of retail outlets was the backbone of your organization, or your phone team was, or your distributors were.

Once you had a set of marketing tools, the organization got in line. You made what you could sell, and you sold what you could market.

A week after I got to college, I signed up to be a sales rep for the *Harvest Journal*, a magazine you've never heard of. Though it was only 1978, the editors were publishing a New Age magazine, a literary journal about a new era. My foggy recollection is that the articles were pretty good, and the printing quality was excellent. My job was to ride the Boston subway system, stopping at just about every stop and selling ads, door to door. I wasn't just a sales rep. I was pretty much the entire marketing department.

Five months later, I hadn't sold a single ad. (I almost sold a one-eighth-page ad to Buzzy's Sirloin Pit. I was so desperate, I offered to rebate my commission just so I could have a sale.) Because I didn't sell any ads, the magazine folded.

Obviously.

Wait. Not so obvious. The magazine didn't fold because the editorial staff couldn't write good articles or because people didn't want to read it. It folded because the real target customers (the advertisers) weren't persuaded by the marketing (me) to support the magazine. The *Harvest Journal* was out of sync with its market (the advertisers) and the method of reaching that market (me).

We make what people buy. If it doesn't sell, it doesn't get made.

A year or two later, I started a snack service. The idea: Hire college students to walk up and down the corridors of the dorms, yodeling "Bagels," pronouncing the word as if it had four or five syllables. For a dollar, you got a bagel and a little foil package of cream cheese. We bought our supplies (in big black-plastic garbage bags) for about ten cents each and paid the students a commission of a quarter for every bagel sold.

We sold a lot of bagels. Tens of thousands of them.

Was it the bagels? Of course not. They were ordinary bagels. What made the business work was the alignment of our customers, our sales force, and our distribution method. Nine P.M. at Tufts University was not a good time to get hungry—at least not until we came along.

You have probably heard this before, but perhaps never in quite this way: Successful organizations are built around successful marketing tactics. Without the tactics, there's nothing.

The One Big Insight

While I hope that there are a bunch of interesting ideas inside this book, there's exactly one big insight. You just read it, but it's worth repeating:

New Marketing favors some approaches over others.

Some tactics, products, services, organizations, distribution models, and stories work better with New Marketing than others do.

So what?

So why would you want to use the wrong tactics with the right techniques? Or vice versa?

Avoid compromise. In a world of choice, no one picks something that is good enough. In a world of networks, few pick the isolated. In a transparent world, people avoid the deceitful.

If the organizations of yesterday were optimized around the advertising, selling, and retail tools of that era, what happens now?

The New Marketing doesn't demand better marketing. It demands better products, better services, and better organizations.

Is It All About the Yodeling?

Here's what happened while you were busy doing your job: The basic tactics of spreading ideas about your products (yodeling college students, billboards, radio ads, and commissioned salespeople) started to evolve. Slowly at first, but then faster and faster. Innovations have not only created new ways to spread ideas, but have also made some of the old ways far less effective.

At the same time that there are whole new ways to market, the tactics of building a business haven't caught up yet. Marketers are trying to play a new game, but their co-workers are still busy playing the old one.

If marketing is at the core of every organization, and marketing is different, then the organization that surrounds it must respond. Marketing doesn't support the organization. The organization supports marketing.

Nearly ten years ago, in 1999, Procter & Gamble made a huge bet on customized cosmetics. The idea behind Reflect, its line of makeup, was that any woman could formulate her own specific kind of makeup. P&G offered more than fifty thousand combinations of lotions and potions for every complexion. P&G had sneak previews and offered regular users free trials.

In order to make Reflect work, P&G built an extraordinary multimillion-dollar factory. It was capable of turning out thousands of cases of cosmetics a day, each customized via a direct connection to the user via the Internet.

In 2005, after six years of losses, P&G shut down Reflect.

What went wrong? Certainly there were plenty of women online, and more every day. And online shopping was all the rage, with billions spent on everything from handbags to books.

Reflect failed for a simple reason. It failed because it was run by P&G. America's greatest marketer, the king of advertising and in-store displays, tried to build a mass-market business on top of a micromarket platform. They overstaffed, overplanned, overbuilt, and demanded that consumers respond in a way they expected. P&G jumped in with both feet, swallowing the hype of the moment, and it cost them a fortune.

Compare P&G's experience to Ralph Lauren's. Lauren, a major provider to upscale department stores, discovered the outlet mall in the 1980s. Just about everything that Ralph Lauren had built to be successful in an upscale retail store was exactly wrong for succeeding in an outlet store.

- They didn't have enough outlet-style inventory to deal with the high turnover of an outlet mall. They would surely sell out of damaged or mislabeled goods in no time.
- Their product quality was too good to make a profit at the prices that outlet shoppers demanded.
- The fixtures and personnel training that consumers had come to expect from Ralph Lauren were prohibitively expensive and unjustifiable in an outlet mall in rural New York or Pennsylvania.

While these hurdles would have scared away most manufacturers, Lauren did something smart. He reinvented what he was making to match the retail tactics that would work. So, the Ralph Lauren Outlet Stores:

- Don't sell all that many leftovers. The clothes are almost all made just for the outlet stores.
- Don't sell highest-quality items. Instead, there's an entirely different line than the stuff made for Bloomingdale's.
- Have fixtures that are clean but industrial and cheap. The staff are quickly screened and trained. They don't pretend it's a real Ralph Lauren store. They just sell stuff.

As a result of selectively building stores and running them the way the market demands, Lauren has built a second stream of revenue that is big enough to have a significant (perhaps dominant) impact on the core business.

Building the Foundation: Going Deep

Imagine a simple diagram that represents the functions of an organization:

Close sales.

Make noise about what you sell.

Determine how to distribute what you sell.

Gain an advantage through manufacturing, technology, or research.

Every successful career and every successful organization has a foundation based on some business advantage. This foundation supports a distribution strategy, which is topped by a marketing and sales strategy. The flashy part is the top. That's where traditional marketing comes in. The top of the structure is about getting people to do business with you, to leverage the foundation at the bottom. Many organizations were built with the premise that the noisy, flashy part on top would pay for the rest of the operation. The factory makes stuff and hands it off, and someone else sells it.

Traditionally, when marketing and sales fail to deliver the growth that management wants, well, that's because marketing and sales did a bad job. The job of marketing is to leverage the base, to spend the company's hard-earned profits to spread the word to people who are targeted to hear it—people who will buy. Change the noise at the top, but don't mess with the foundation of our organization. "That's our anchor," says the boss.

So Microsoft's business is anchored in its Windows monopoly. Detroit's business is anchored in its ability to make large numbers of average cars for the middle of the market. The accounting firm down the street is anchored in its legions of hard-working accountants, following a rule book and marking up their time. In each case, this foundation has been honed and refined over time.

The job of marketing is to take that advantage, that anchor, and turn a profit by selling the idea that supports the anchor. You go to Harvard, pay your dues, and use your résumé to get a return on your investment in the form of a diploma. An entrepreneur builds a factory, invests in people and machines and training, and then uses marketing and a sales force to turn that effort into more sales.

The assets at the bottom were aligned with those at the top. That Harvard degree is the foundation that allows your résumé to get you a job at Goldman Sachs. That widget factory, with its high-quality, standardized operations, is the foundation that allows your marketing team to sell widgets nationwide. Sell average stuff to average people, and do it with enough volume, and you thrive.

This has been our success formula for a hundred years, but now, traditional marketing and sales strategies appear to give us less growth than we need. So marketing gets pressured—pressured to use *this* blog or *that* tactic to get more people to buy what the organization is anchored on. The number one question every New Media person gets asked is, "How can I use the New Marketing to help me sell more of what I'm already working on?"

I want to be clear here: we *need* meatballs. We need factories. We need average stuff for average people. What's changed is that meatballs and factories are no longer sufficient to deliver the growth that investors demand. The factory that's already in place makes meatballs. The old kind of one-size-fits-all interruption marketing supported the old kind of organization and vice versa. There was alignment. The organization thrived. But what happens when people don't respond to the old messages any more? What happens when marketing changes and suddenly the rules are different?

If the factory makes meatballs, and the marketing now supports sundaes, you've got troubles (not to mention indigestion).

The problem is that almost everyone affected by the death of the Old Marketing is busy playing with the top of the ice cream sundae that is the New Marketing. The art of marketing has changed, but in too many cases, the base of the organization has not. The efforts put into online shopping, blogging, search-engine optimization, voice-mail systems, and the like tend to be about rearranging the top of the sundae—making more noise and being a little flashier.

While everyone on your marketing team is busy making more noise, the foundation of your organization doesn't match the noise. New Marketing is exciting, but it seems to work best for people who start from scratch, who build their foundation around the idea of their marketing. The very advantages our organizations are built upon are fading, and no amount of flash is going to help sell these meatballs.

The good news is that you don't have to start from scratch. In fact, existing organizations have significant advantages. The challenge is that you have to be willing to become an organization that is in sync with the New Marketing.

Driving Innovation Deep

Here's another look at the pyramid of the organization, with annotation:

CLOSE SALES

The challenge of closing sales never changes. Human beings hate to make commitments, because commitments represent risk, and risk is frightening.

As long as there are organizations and businesses and stores that need to make things happen, there will be sales to close.

MAKE NOISE ABOUT WHAT YOU SELL

This is why everyone is noticing the New Marketing—it's because the noisemaking has changed so dramatically. We wonder about advertising on YouTube, we wonder about how Google AdWords works, and we wonder about creating viral marketing campaigns.

Making noise has always been the fun part of marketing, but making noise is not where the highest returns lie.

DETERMINE HOW TO DISTRIBUTE WHAT YOU SELL

At the same time that the noisemaking was being revolutionized, so was distribution. Not just online sales, but person-to-person sales, megastores, microstores, subscription programs—from one end of the spectrum to the other, the way things are distributed and sold is changing.

CafePress.com sells millions of dollars' worth of imprinted items every month. And they do it without a store of any kind. Instead, they have hundreds of thousands of people, each running their own stores with their own inventory.

When Lynn Sulpy chooses to sell her dramatic oil paintings on etsy.com instead of relying on a gallery, she's making a commitment to something bigger than a noisy marketing campaign. It's a transformation not only in the way she brings products to market but also in the way she decides what to paint next. She doesn't paint to please a gallery owner or a critic. She paints to please herself and to please the clients who are attracted to her vision.

Audible.com offers far more books on tape than any library or bookstore ever could. None of them are sold by Audible in a physical store. By taking advantage of the infinite shelf space of the Web, they change the structure of an entire industry.

These new forms of distribution aren't limited to arts, crafts, and literature, though. You'll find farmers (The Chef's Garden, in Huron, Ohio, sells more than six hundred types of obscure vegetables, straight to restaurants), manufacturers (Cut-Mark, which specializes in bushings under 1⅝" in size), and nonprofits (like roomtoread.org, which does no traditional fundraising) using the same techniques.

GAIN AN ADVANTAGE THROUGH MANUFACTURING, TECHNOLOGY, OR RESEARCH

Here at the base of the pyramid is where the real transformation occurs. Just as the assembly line revolutionized the idea of a factory, the New Marketing demands a transition to a different way of thinking about products and services and how they are invented, designed, and produced.

The New Marketing has exposed great work to large numbers of people. Those ideas spread farther and faster than they ever have before. While this is a nice perk to people who make something a little better than ordinary, it's a huge revelation to those who are able to make the extraordinary.

What's happening? You can now be rewarded very quickly if you can build something worth talking about. The distance between the brain of the designer and the ear of the consumer is now much, much shorter than it ever was before.

Old Marketing and New

Old Marketing is the kind we grew up with. Old Marketing is TV ads for Trouble,® with Pop-O-Matic.® Old Marketing is a command-and-control approach to the creation and spread of ideas.

New Marketing is about fashion and stories and permission and promises. New Marketing doesn't understand top-down command-and-control thinking. It's actually caveman marketing, the sort of marketing that existed before money and corporations took over.

See the handy table on page 26 to help you tell them apart.

OLD MARKETING	NEW MARKETING
Limited number of media outlets	Countless media outlets
Limited physical retail outlets	Countless online retail outlets
Emphasis on horizontal success (hits)	Emphasis on vertical success (niches)
Marketer-to-consumer communication	Consumer-to-consumer communication
Barrier between consumers and makers	Permeability between consumers and makers
Spam	Permission
Product line limited by factory	Product line limited by imagination
Long product cycles	Fads
Market share	Fashion
Features	Stories
Advertising a major expense	Innovation a major expense
Large overhead = stability	Small overhead = low risk
Customer support	Community support
Focus groups	Launch and learn

The Difference

I attended an all-day brainstorming session with one of the oldest, best-known nonprofits in the country. They have a fancy Web site, loaded with Flash features, tell-a-friend buttons, and a blog.

Last year, the site raised two million dollars. This year they want to do more.

With a mailing list of five hundred thousand e-mail accounts, this organization has demonstrated that they can extract money from people who sign up for "e-mail blasts." And the stated goal of the group is to increase the size of the list by a factor of six, to three million. Then, using free stamps (e-mail), they can hammer this list to raise a lot of money for their good work.

Compare this organization to Kiva. Kiva is a brand-new organization that, after just a few months, generated nearly ten times as much traffic as the older group. And they are raising more in a month online than competition does in a year.

Is it because they have a better site?

Nope. It's because they have a different sort of organization. They created a Web-based nonprofit that could never even exist without the New Marketing. One group uses the Web to advance its old agenda, while the other group is of and by and for the Web.

One is focused on market share, on getting big by controlling the conversation. The other is into fashion, in creating stories that spread because people want to spread them.

And that's the schism, the fundamental demarcation between the Old and the New.

One organization wants the New Marketing to help it grow a traditional mailing list so it can do fundraising and support a traditional organization.

The other (Kiva) is creating an organization that thrives on the New Marketing rather than fighting it.

Kiva works because the very nature of their organization requires the Web at the same time that their story is so friendly to those who use the Web. Kiva connects funders (that would be you) with individuals in the developing world who can put a microloan to good use. Doing this in a world of stamps is almost impossible to consider. But doing it online plays to the strengths of the medium, and so, at least for now, the users of the medium embrace the story and spread the word.

Please note that I'm not insisting that everyone embrace these new techniques. All I'm arguing for is synchronization. Don't use the tactics of one paradigm and the strategies of another and hope that you'll get the best of both. You won't.

After just a few minutes of conversation at the older nonprofit, one person realized, "So, if we embrace this approach, we don't have to just change our Web site—we're going to have to change everything about our organization. Our mission, our structure, our decision making. . . ." Exactly.

Marketing Enables (and Demands) a New Organization

Organizations used to be built around very expensive marketing tactics—things like real estate or TV commercials. Organizations used to rely on slow, unpredictable communications tactics to reach their employees and prospects. Organizations used to have to embrace slow cycle times and difficult and expensive marketing rollouts.

There are more than ten thousand McDonald's restaurants in the United States, a number driven by a generation that had just discovered the car. The ubiquity of the stores isn't just a real estate investment—it's a marketing one. If I asked you how much McDonald's spends on marketing, you'd hesitate to say "billions of dollars a year," but if we count the cost of the real estate, that would be the right answer.

The Catholic Church has a deliberative, centralized decision-making process that was the state of the art when messages traveled by horseback. By coordinating planning and decision making, they were able to dominate large portions of their marketplace. Today, of course, their response time and apparent inflexibility make it hard to compete with smaller, more nimble groups. The organization of the Catholic Church was driven by Old Marketing tactics. Yes, they were marketing tactics, not organizational tactics. They were marketing tactics because they existed to help the Church grow. And now those tactics have changed. Has the Church?

Aided by technology, the world now acts smaller and works faster. Successful organizations of yesterday built structures that were big and slow and centralized. As the world has digitized, these older, slower systems are getting in the way.

Why didn't American Express invent (or buy) PayPal? Connecting one customer to another is what PayPal did, but American Express wanted to have more control, so they watched the opportunity go by, and eBay bought the company instead. This was a marketing decision from the start. AmEx is based on a top-down, centralized, Old Marketing model. They couldn't see the PayPal opportunity, because the idea of growing by connecting one customer to another makes no sense to them.

Why did *Fast Company* fade away? Not only was it a great magazine, but the conferences and community they created had a huge influence on our society. Everyone who ever attended a RealTime Conference remembers the impact the magazine and its founders had on them. The organization had a chance to build an entire portfolio of assets, but focused on making a paper magazine instead.

When the growth in magazines faded, *Fast Company* got hurt. It never reached its potential, because it was built from the ground up around the Old Marketing model of interrupting subscribers with ads, instead of embracing the idea of connecting subscribers to each other.

Why didn't Barnes & Noble become Amazon?

All or Nothing

Years ago, the folks at Wal-Mart flew me out to Bentonville to give a speech to their online division. I was stunned to see a huge banner on the wall. It said, "YOU CAN'T OUT-AMAZON AMAZON."

A quick look at the Alexaholic traffic comparison chart below shows that Amazon regularly gets thirty times as much Web traffic as the world's largest retailer does.

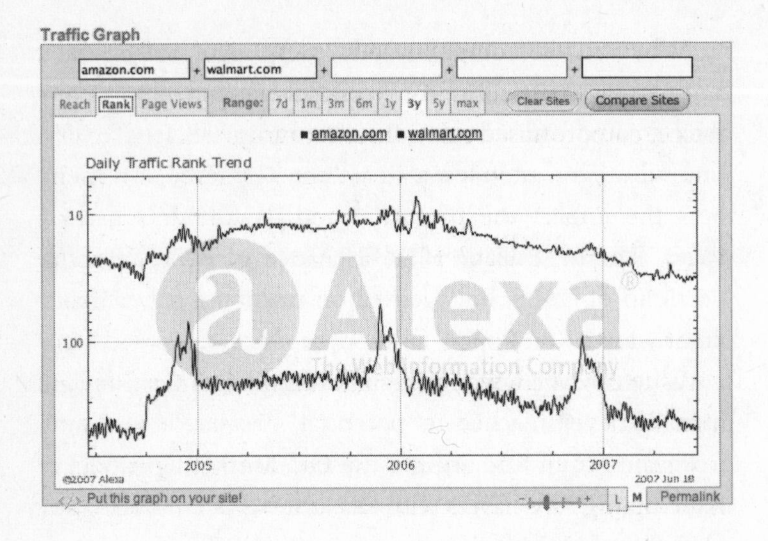

Wal-Mart executives realized that online shoppers had a nearly infinite number of choices, and if Wal-Mart wasn't prepared to be the biggest, best, most useful online shopping outlet, they would fail. So they didn't try.

Sure, they do fine around Christmas each year. But they didn't fall into the trap of half-measures. Which makes it even more clear how silly it is for an independent bookstore or a local pharmacy to try to build an alternative that can compete with Amazon on its terms.

Amazon is the "all." They embraced the medium and won. Many other companies are the "nothing." They delivered a little less, made some thoughtful compromises in order to straddle competing priorities, but failed because their competition refused to compromise. In a world of choice, compromised solutions rarely triumph.

The Man Who Invented Marketing

No, it wasn't Ron Popeil, father of the Ronco Veg-o-Matic. It was Josiah Wedgwood (grandfather of Charles Darwin, actually).

Josiah Wedgwood lived, as we do, during a time of dramatic change. We have it a lot easier, though. In 1730, when Wedgwood was born, it wasn't at all unusual for an Englishman to never travel beyond his village. It wasn't unusual to earn a subsistence living, either by scratching out some food through farming or by working (from childhood) in a mine or small workshop. The average lifespan then was thirty-three years. The ideas we take for granted today—not just central heating or a food market, but doctors and clean clothes and mobility—were completely unknown.

Wedgwood, like many in his family and his village, apprenticed to be a potter. Several decisions, however, led him to become the most successful potter of all time.

First, he saw that he could grow his business. This was a critical component of the first industrial revolution—the idea that a business could be more than a craftsman and a few assistants or apprentices. It's easy to underestimate how difficult this was to see then, but even today most businesses have self-imposed (but unspoken) limits on their growth.

Second, he chose to grow by viewing his product as more than the bare minimum to get the job done. He's famous for smashing imperfect items with his walking stick, proclaiming, "This will not do for Josiah Wedgwood!" He saw that his product was more than just a useful object; it was part of a continuum of output, something that could carry his name to a large number of potential new customers.

Remember, in 1730, it was still common for people to eat their meals out of a bowl of stale bread. Ceramics, a plate of any kind, was a bit of a luxury. The idea that you would smash a plate just because it was a little irregular was nuts.

Wedgwood had perhaps the first important brand name in the history of commerce. Unlike any other potter (or brewer or carpenter or blacksmith), he put his name on every item he made. By organizing his company around a brand, he changed the dynamic of organizational growth.

Wedgwood took marketing much further than that, however. He invented the idea of appealing to the newly monied masses by first selling directly to the upper classes.

Wedgwood risked his entire company in 1771 by investing more than $2.7 million and sending, without asking first, samples of his pottery to one thousand wealthy Germans.

Believe it or not, more than half of them turned around and ordered more pottery. Several years before that, Wedgwood gave Queen Charlotte (the wife of George III) a breakfast set. A few years later, she ordered a full tea service, which of course he turned into Queensware™ and sold to the masses. He also created bespoke china for Catherine the Great of Russia. Though he made a small profit on the commission, his real profit was in displaying the finished works for months in London before shipping it off to Russia. That much-talked-about display attracted hordes of people to his showrooms (another innovation) in London. And many of those visitors turned into customers. (Did I mention that his sales force worked on commission?—another breakthrough.)

While many of these innovations appear to be flashy marketing tactics, none of them would have been possible without the wholesale overhaul of the way his pottery was made. The factories had to be designed and created in order to deal with the demand, and just as important, the technology for making pottery like no other had to be in place as well.

He dealt with increased demand by creating one of the first assembly lines. Each worker mastered a particular stage in the process, and Wedgwood and his managers were frequently reorganizing the process to increase productivity. As part of his overhaul of the production process, he developed one of the first time-clock systems.

What Wedgwood did that was so extraordinary is that he redefined, from the ground up, what it meant to be in business. He practically invented the ideas of standardized quality control, of innovative design for utilitarian products (his cauliflower teapot was a sensation), and even of health insurance (of a sort) for his workforce. Wedgwood understood, long before his peers in other industries, that a new class of customer and a new kind of distribution enabled a new kind of organization to thrive.

Wedgwood didn't hire potters or people who had apprenticed with potters. Instead, he hired the untrained, and he trained them himself. He realized that his way was the new way, and that it was easier to teach someone than to unteach them first. Wedgwood's pottery sold for four times as much as that made by smaller potters in the same region of England—because he made it and marketed it differently.

It's worth noting that the Wedgwood factory was built directly on the Trent and Mersey Canal, a seventy-seven-mile-long waterway that Wedgwood lobbied Parliament to have built. Why? Without a way to get his wares to market, he had no product. Taking control over every element of his product—from the way it was made to the way it was delivered to the stores that displayed it—Wedgwood created the environment he needed to thrive.

Was Wedgwood a marketer? Only if we realize that marketing demands an organization that matches it. By taking advantage of changes in the environment, Wedgwood was able to change his organization in response. Completing the cycle, the environment changed in response to his new organization. This virtuous cycle enabled him to die with $44 million in his estate.

Following in his footsteps was every other industry that we know today. Those organizations in those industries eventually morphed to become the company you work for right now. And as we enter a new era, we need new Josiah Wedgwoods to reinvent those companies (your company) for a new age.

Marketing Advice for Thomas Wedgwood

Josiah served his apprenticeship with his older brother Thomas. Thomas was a typical potter of his day, practicing his craft in a small workshop, turning out typical pottery, and selling it at commodity pricing. Thomas is forgotten today, largely because he ignored the marketing advances of his day—he insisted on running his business the way it had always been run.

Here's the marketing advice that Josiah might have given Thomas:

- Create new ways of glazing your goods and novel items that people seek out.
- Sign your work.
- Increase your pricing by 400 percent.
- Establish high quality standards.
- Build a bigger factory and put it near a canal, one that you lobbied to have built.
- Train nonpotters to join your workforce, and design innovative ways to manage them.
- Organize the factory to reflect a separation of labor.
- Open showrooms in London and change the stock weekly.
- Focus on mass production.
- Sell to the richest people in the world, but avoid taking custom work except for items for heads of state.

Most marketers wouldn't consider all of this to be marketing advice. It feels like management advice. Exactly. Josiah

Wedgwood probably gave this advice to his brother, an indigent potter with one or two apprentices. Thomas didn't try to do much of anything. It probably felt like trying to walk to the moon.

And now, almost 250 years later, we're in exactly the same spot again.

Some organizations are going to watch in disbelief as the world changes. Others, though, perhaps inspired by Josiah Wedgwood's fortune, will start walking.

Avoiding the Meatball Sundae

The idea of synchronizing your organization with the New Marketing isn't brand-new. In fact, it's already working for someone else, perhaps your competition.

Some organizations, mostly new ones, have figured out the power of alignment. A blogger in California is making a million dollars a year writing his blog. Not because his blog is better than anyone else's but because his organization and revenue model match his marketing. Most for-profit blogs are extensions of other for-profit businesses and exist to support those businesses. As a result, they make trade-offs in order to satisfy the mother ship. This blog, on the other hand, embraces its medium and does just one thing . . . without compromise.

A fourteen-person hedge fund upstairs from my office is making a few million dollars (per person, per year), staffed with people working just minutes from home. By realizing that a very quick, extremely talented group of people could profit without being on Wall Street, they built an organization that matched the tools available to them. On the second floor, between my office and that hedge fund, is a fast-growing firm that can translate just about any text into just about any language . . . and they do it all over the world. One more time—the entire organization thrives because they built themselves around a different kind of marketing.

And it's not just small companies that are using the tools of the New Marketing to do business differently. Chart-topping rappers show up on YouTube as often as they do on MTV. And the Republican Party is investing millions in creating a grassroots coordination ability designed to last a generation.

Smart organizations are investing time and energy into transforming their assets. They know the New Marketing is more than a hot topping. Instead they use New Marketing to dig deep, to redefine what they actually do to add value. The new rules are here and they're not going away. If your assets are synchronized with what you can do with the New Marketing, you win.

Old (Good) vs. New (Wacky?)

My old boss at Yahoo! would label any nontraditional idea as "a little wacky." Rather than exploring the edges of what the Web could become, he was obsessively focused on the tried and true.

The reality of traditional marketing is well entrenched. We've come to expect that retail stores are open on Sunday, that designer labels sell for more than store brands, that persistent salespeople do better than nonpersistent ones, and that big companies have receptionists and handsome lobbies.

You've got a job to do, an organization to market. You don't have time for the tech-heavy jargon and rapidly changing rules of the Web. You need to sell more widgets, get more people to use your service, or increase donations to your nonprofit. You don't have time for YAA (yet another acronym), and you don't have the money to spend on the servers and programmers it will take to stay on the leading edge. They don't call it the "bleeding edge" for nothing.

So, like most organizations, you've taken a two-pronged approach. The first prong is "real" marketing. That's the stuff you spend big money on, the core of your efforts—TV, magazines, direct mail, a sales force. The second prong is "experimental" marketing, the new stuff, the stuff you don't really expect to work very well. (And it's not an accident that your entire organization is aligned behind the first kind, the kind that actually works.)

Even companies like Apple and Microsoft make a distinction between "real" and New Marketing. They put their big dollars and their most senior people on stuff like publicity, packaging, retail relations, TV ads, magazine ads, and pricing. Then, there's a different group, the young turks—they're the ones who worry about New Marketing.

I'm arguing that something just shifted—that the New Marketing isn't just about technology, is not just an online phenomenon, and isn't wacky. Not anymore.

The Fourth Industrial Revolution

According to Thomas McCraw, author of *Creating Modern Capitalism*, there have been three industrial revolutions.

The first one occurred from 1760 to 1840. This was the giant breaking point, the moment our world shifted from tiny individual-based businesses and family (or feudal) farms to a world we'd be more likely to recognize today. Before this, even the idea of an employee was strange.

Steam engines and water power made production possible. Available capital meant that entrepreneurs could borrow money and put it to work. And available labor meant an organization could have as many as a few hundred employees, leveraging the machines that made it possible to produce textiles, pottery, and iron. Josiah Wedgwood was one of the poster-child CEOs of the era.

The second revolution is the one we think of when Henry Ford comes to mind. From 1840 to 1950, factories got more efficient. More employees were available, and they were better trained. Capital was cheaper, so entrepreneurs could invest more. Electricity and assembly lines and roads and trucks meant that a huge range of items could be produced and delivered around the region, if not the world.

The third revolution was led by mass marketing and newly enriched consumers. Now that employees were getting paid more, they could buy more. Factories focused on items consumers wanted, not needed. Mass marketing ensured demand, and companies organized themselves to produce for the masses. Information became critical, and so did being the best in your field.

Coordination and communication were the key components of the third revolution. Suppliers could contact each other quickly and cheaply. Retailers could communicate with consumers and suppliers. New products were derived from older products, and companies became intricately coordinated. Whereas Henry Ford made every single part of a Model T in his own factory (at one time, Ford employed his own shepherds so that he would have a supply of wool for the seats), the typical Toyota car of this era was built with parts made by dozens of other companies. Intricate relationships with suppliers became the norm, not the exception.

By the end of this biggest of the three revolutions, more than half the people in the industrialized world made their living without being industrial—the service sector dominated. Services—connecting people and ideas with things that are being made—were the cornerstone of this era.

And now we enter the fourth revolution. Just ten years after the birth of the Web, New Marketing has so fundamentally changed the dynamic of production and growth that the rules of the third revolution are no longer dominant. Instead, the trends of New Marketing require a new kind of organization and a new way of doing business.

The Point of No Turning Back

New Marketing isn't about technology any more than fast food (and the drive-through window) is about cars. Technology, most especially the Internet, has enabled the New Marketing, but you don't have to understand it to use it.

It's been a decade since the underpinnings of this revolution were put in place. For the last ten years, ignoring most of what was going on with the New Marketing wouldn't have hurt you that much.

Unless you were a travel agent. Now you're out of business.

Unless you worked for a newspaper. Now you're out of a job.

Unless you bought books. Now you know that the best place to find any title, fast and cheap, is online.

Unless you sold insurance. Now you've seen that fat commissions are a thing of the past.

I could go on, of course. I could talk about how the millions of full- and part-time real-estate brokers are facing an uncertain future. How everyone from chiropractors to workers'-comp agencies are facing a challenge in getting new business. But I won't.

Instead, I'll remind you of the tens of thousands of organizations that didn't ignore the new rules. These are the organizations that have grown, because instead of dabbling, they've embraced a dozen or so simple principles and driven them deep into the structure and DNA of their organizations.

- 8CR, the independent bookstore that is now the most influential seller of business books in the country

- Threadless, the two entrepreneurs who built a T-shirt business that does more than $20 million a year in revenue—and doesn't even have an artist on staff
- The lawyer who built Techcrunch, a blog that generates a million dollars a year in revenue
- The car dealership in Syracuse that happily sells cars to people who live hundreds of miles away in New York City
- Kiva, the nonprofit organization that raises money in Kansas and funds craftspeople in India . . . without an office in either place

The list goes on and on. All of these organizations are built around fourteen trends that show no sign of going away. Your challenge is to ride the trends, to ride them in an organization that's designed to run with the New Marketing, not fight it. Not partly, not in half-measures, but all the way. If Josiah Wedgwood were alive today, he'd be saying, "Let me out of this box. It's dark! It's dark!" Sorry. I meant, if Josiah Wedgwood were alive today, he would rebuild every one of his factories and get his organization in sync with the realities of the New Marketing.

I want to share a list of trends that have fundamentally changed the dynamic of running and growing an organization. This is not a complete list, not by any means. Nor does every item on this list affect every organization.

Just as Josiah Wedgwood profited from having a very different vision of the world than his brother Thomas did, understanding how these trends will remake your industry is the unfair advantage that will enable you to get in sync and thrive during the next revolution.

The Old List

My list of trends follows. It's tempting to compare this list to the old list (TV works, mass is king), but that's not fair. It's not fair because the old list had only two items on it. Together, these two principles—the idea that you can profitably interrupt consumers with ads and the idea that creating average stuff for average people at low prices is the way to defeat the competition—drove everything.

My list has more than twelve principles. Which means that no trend by itself is enough. Sam Walton built Wal-Mart on the back of just one. Today's successes are the result of some of these trends in combination. More than anything, today's successes embrace these trends just as completely as Procter & Gamble embraced the old list a century ago.

Marketing has always been about discovering what people want and need, and telling them a story about how they can get it (from you).

It's easy to jump to the conclusion that the list represents a change in what people want or need. It doesn't. Human nature hasn't changed a bit. What has changed is the environment we live in. The combination of technology and competition has led to a world where many people can get what they want, when they want it. A world where people have control over the attention they give to marketers. A world where we have so much income, so many assets, that we can demand just about anything we can imagine.

So, as you review these trends, realize that the most important thing that has changed is the ability of consumers to finally have what they've wanted all along. And no, that's not a 5-percent discount on a muumuu at TK Maxx. What we've wanted all along is to be treated with respect and to be connected to other people.

Part 2

THE FOURTEEN TRENDS

TREND 1

DIRECT COMMUNICATION AND COMMERCE BETWEEN PRODUCERS AND CONSUMERS

Organizations hear more, and more often, directly from consumers. Organizations can also sell directly to customers, eliminating the middlemen. And they can build a permission asset, which allows them to market directly to prospects. Even better, organizations can create products for their customers instead of searching for customers for their products.

The Need for Speed

In the real world, a retailer understands that making customers wait leads to lost opportunities. Even doctors have discovered that in a competitive environment you can't require patients to sit for long in the waiting room. The online world has compounded the need for speed. Just ask Brooke Kaltsas. She prices jobs for Webstickers.com, a Web site that, no surprise, prints stickers.

If you need stickers, you go to their site, pick out what you want, and then submit the job to them for pricing. I posted a job Sunday evening at nine. I heard back from Brooke at twelve thirty P.M. on Monday.

Anywhere else in the printing industry, that sort of pricing turnaround would be considered astonishing. After only three business hours had passed, she came back with a price.

Too slow.

Too slow by an hour and a half. Bandmonster.com had closed the sale two hours earlier.

And only because no site I found in a few minutes of searching would do the job without making me wait even that long.

Companies accustomed to relying on their inbox as a time buffer have discovered that this just doesn't work any longer. You used to be able to wait a few days . . . or even a few weeks . . . to get back to a prospect or to resolve an open issue. Now, in the era of one-click shopping, consumers have been trained to expect that they should be able to work directly with the person who can make something happen, and that it should happen immediately. The promise is now implicit: We will get back to you within moments, or you should go elsewhere.

Inbound communications from consumers demand speed.

No Insulation

Sonos understood that it had to rip down the walls.

Sonos makes a device that would have been inconceivable only ten years ago. The Sonos stereo system is a remote control (with an LCD screen), a hard drive, and a box. The box hooks up to your stereo speakers, and the hard drive holds all your MP3 files. You can use the remote to review your entire music collection and play it anywhere in your house. Add more boxes, add more rooms. One hard drive can be used to let your daughter play Mahler in her room while you listen to Coldplay in the kitchen.

While the idea is simple, and installation is a snap, the products Sonos replaces weren't simple or easy. As a result, the previous generation of multiroom speaker systems were sold by consultants—the sort of private services that cater to multimillionaires and their homes. At a recent CEDIA Show (the conference for installers), one of the categories was "Best Installation over £100,000."

If you were in that business (companies like Runco and Stewart Filmscreen), you catered to the CEDIA installers. You let them be the middlemen, the service and support people, the installers, and yes, the folks who made most of the profit. The installers guided many of the decisions that their clients made, and you were at their mercy.

Sonos sells a product for about a thousand dollars. That's less than the gratuity on most custom installations. As a result, Sonos decided to use the Web to allow consumers to interact with them directly.

In addition to a well-designed discussion board, Sonos invested in motivated, well-trained online staff members, who are seemingly everywhere, answering questions within a few minutes of them being asked. Sonos has pleasant technicians answering the phone on weekends. They not only publish their e-mail address but actually answer queries (and helpfully) in a matter of hours.

Of course, Sonos is still happy to work with the CEDIA crowd. But by embracing the ideas of accessibility and speed, they have made their product appealing to people who could easily afford to spend ten times as much.

How will their competition catch up? How can their competition simultaneously jettison their entire sales force, dramatically increase the quality of their customer service, lower their end pricing by 70 percent, and make the product consumer friendly? They can't.

When everyone was playing by the same rules, when all suppliers relied on insulation in order to maintain margins and keep throughput efficient, it was a terrific system (for the sellers, anyway). But as soon as one player in the industry can use a direct connection to the end consumer, the rules change for everyone.

The Deluge

The fact cannot be denied: Your people (customers, employees, prospects, readers, whatever) want to be heard. They demand to be heard.

The volume of letters to the editor more than doubled in less than a decade at the *Raleigh-Durham News & Observer*. While newspapers are seeing more and more mail, they're actually publishing the same number of letters . . . which means readers are going online to see what others are saying. The newspaper isn't the watercooler anymore, not when there's a blog on your favorite topic with dozens or hundreds of ideas exchanged every day.

The U.S. House and Senate have seen increases of 50 percent, year after year, in e-mail volume to their offices. In 2002, when they gave up counting, the typical congressperson was getting more than five hundred e-mails every day. Today it's probably ten or twenty times that amount.

The irony is that most organizations (and politicians) view this as a problem, not as a spectacular windfall.

When I Send a Note to Your CEO, Who Gets It?

A lifetime ago, the *New Yorker* ran an article about an e-mail correspondence with billg@microsoft.com. Apparently the world's richest entrepreneur was busy answering his own e-mail. In the wake of the article, naturally, that went away.

While you have no idea if the mail will get read, it's possible to send mail to Bill Gates, to the president of the United States, to the head of a big ad agency, and to Donald Trump. And then what happens?

An inbound e-mail is not (just) an expense; it's (also) an opportunity—a chance for your organization to eliminate barriers and have a dialogue with a prospect or a customer. The logical place for a person to start the relationship is with your CEO. She's the famous one; she's the one on the cover of the magazine.

What happens then?

In most organizations, absolutely nothing. No response of any kind.

After hearing some of the original hype about its new MP3 player, Alex McAulay tried to get ahold of someone from Microsoft to market the Zune on campus at his university.

"Hi, my name is Alex McAulay, I'm calling on behalf of the business department of the University College of the Fraser Valley. The school has ten thousand students. I want to partner with Microsoft to help release new products through word-of-mouth marketing on campus. Can I please be directed to someone who can help me start this?"

He spent three hours getting bounced all over Microsoft. Telephone tree terrors. Microsoft made him sign up to become a partner, a process that took another hour. Then he spent another ninety minutes on the phone getting bounced around.

The Zune didn't get promoted at Fraser Valley this year.

You know all the reasons for the screen between consumers and organizations. Inbound correspondence is a hassle, it's expensive, it leads to never-ending dialogue, and you're worried about getting sued. But what is it worth?

Accenture is a huge organization with hundreds of thousands of employees. If you were a prospect, or perhaps a dissatisfied customer, where would you turn? Don't try reaching the CEO, Bill Green. The company Web site has his picture, but not his phone number, his e-mail address, or his snail-mail address.

I don't think consumers get the runaround because of a deliberate plan by the organization. Instead, it's based on history. A generation ago, the only people worth talking to had visible power. The big retailers or law firms or manufacturers or media companies were easy to discern, so you knew which calls to take (and people knew where to find you). Today, though, everyone has power (at least a little), and everyone has the ability to make a difference to your organization. So your filters are screwed up. You can't tell in advance who's who.

Bill Green is missing an opportunity every day. Every time he runs a billboard in an airport or has a visitor to his Web site or receives a résumé from a talented person, he has a chance to develop an e-mail dialogue with someone important. By hiding, he may feel like it is easier to deal with the onslaught of inbound information, but he also loses the chance to start a conversation.

What's an E-mail Correspondence Worth to You and Your Organization?

A typical direct-mail campaign earns a 1-percent response rate. For every one hundred letters sent (at a cost of up to a dollar each), the company gets one order.

An e-mail campaign on the other hand—not spam, but real, opt-in permission-based e-mail—can get twenty or thirty times the response of direct mail. And the cost of stamps is gone. In other words, a valid e-mail relationship can be worth two thousand times what a mailing address is worth.

Here's the math:

Cost of renting 1,000 names (cpm), for one-time use: $75
Cost of the mailing, at $1 each: $1,000
Orders received: 10 (1 percent, which is pretty high)
Profit per order: $120
Total Gross Profit: $1,200
Costs: $1,075
Net: $125

versus:

Cost of acquiring 1,000 names via opt-in newsletters, phone calls, inquiries, and sales force: $5,000
Cost of mailing: $0
Percentage that convert to customers in the first month (it decreases each month for the next few months): 5 percent
Total by the time you're done: 20 percent
Number of sales: 200
Profit per sale: $120
Total Gross Profit: $24,000
Costs: $5,000
Net: $19,000

And yet for every company that recognizes the value of an inbound permission, there are a hundred that view it as a nuisance.

Why It's a Nuisance

If the posture of the organization is, "We have enough to do when it comes to our jobs, and now we have to figure out how to deal with all these interruptions," then it's no wonder that inbound correspondence is a problem. If you've got a call center focused on relieving the "problem" and you have management working to cut costs at the call center, it's no wonder all that consumer attention is seen as a problem.

There are thousands of motivated consumers and voters who are dying to interact with you. The challenge is in organizing so that this phenomenon is seen as what it is: a profit center.

When organized sports teams first became commercial, it was easy to see snacks as a hassle. You had to get all that food in and clean up afterward. It took a generation or two for the teams to realize that the game was really the hassle—they were in the food business. (Pacific Bell stadium sells garlic fries, peanuts, and calamari salad. They also offer sushi, soup in a sourdough bowl, kraut dogs, Portobello burgers, steamed mussels, and, of course, popcorn.)

For many organizations, interaction with the outside world quickly becomes their most profitable opportunity.

The Threadless Story

Threadless.com is a T-shirt silkscreening company in Chicago. Founded in 2001, they've seen their sales quadruple every year. In 2005, they sold roughly six million dollars' worth of T-shirts, with 2007 on track to continue the pace.

It's easy to imagine a T-shirt-printing company as focused on one of two things: printing shirts fast and cheap or promoting shirts like mad, with licenses, retail deals, or aggressive advertising. Threadless does neither.

Instead, they sell their $15 shirts (which cost far less than $4 to make) only on their Web site, only to people who came to the site without benefit of advertising. How? By printing designs created by their customers. Every person who visits the site isn't just a prospective buyer. They're also a prospective designer.

Threadless regularly holds design contests. The winning designs win cash. All designs are available for purchase. Nascent designers promote their designs to friends, and some of those friends end up buying shirts. It's a pyramid scheme of T-shirts, one where everyone wins.

You can bet that when you contact Threadless, your e-mail gets read. You can bet that when you make a suggestion, it gets considered. And you can bet that a real human is paying attention. Because paying attention to inbound messaging is exactly what Threadless does for a living.

Permission Math

Let's take this a step further. The *cpm* of traditional spam-based advertising is the cost of reaching one thousand people. Of course, you're not really reaching them. You're not reaching them, because they're working overtime to ignore you. They skip your commercial when they watch with a Tivo, they click away from your ad banner when it's online, and they can't quite remember that superclever billboard you paid so much money for.

It's not unusual for effective advertising media to sell for a $50 cpm. Plenty of studies show that 7 percent is a high number for the percentage of consumers who can even **remember** your ad, never mind actually take action because they saw it.

Which means that you're actually paying closer to $750 for every thousand people who see your ad. (Because only one out of fifteen is actually watching.)

And then, next week, when you want to do it again, you've got to pay another $750. And on and on, forever. No real asset built here. Certainly not a measurable asset that turns into a profit.

The alternative is to abandon spam-based advertising and instead create a permission asset. A permission asset that's carefully built and maintained is often enough to structure an entire company around.

Knock, Knock

That's the reason you got hired. To knock on doors. To interrupt strangers, to spread your ideas, to close sales. The job of marketing is to grow the organization, and growth comes, obviously, from putting yourself in front of people who didn't know about you before you got there.

Which means advertising.

Newspaper ads, TV ads, billboards. The Yellow Pages. Whatever it takes to get the word out.

The entire cost structure of organizations that do marketing is built around this idea. Running for President costs more than a hundred million dollars—but that's not the cost of the staff or even the plane. Instead, the vast majority of the cash goes to interrupting people.

A nonprofit might spend as much as eighty cents on every dollar raised, just to raise more money—to send more mail and make more phone calls.

When ads start to falter, marketers just make the ads louder. We put them in more places. (Urinal ads, anyone? Universal Pictures is now placing ads there.)

How do marketers justify this insane amount of waste? How do we rationalize doing direct-mail campaigns that have a 99-percent failure rate? For a long time, it appeared to be the only option. If you wanted to get the word out, you spent money on media.

No longer. Now, if it's important enough to spend money on, it's important enough to market in a very different way.

Here's the sea change: You have the chance to go from finding customers for your products (the meatball way) to a new way—finding products for your customers. The simplest example of this is the difference between a book publisher (who always seeks new readers for his new writers) and a magazine publisher (who commissions articles for his existing readers). A magazine makes far more money than a typical book publisher because of this difference.

Reaching the Unreachable

Marketing, I think, can be divided into two eras.

The first, the biggest, the baddest, and the most impressive was the era in which marketers were able to reach the unreachable. Ads could be used to interrupt people who weren't intending to hear from you. PR could be used to get a story to show up on Oprah or in the paper, reaching people who weren't seeking you out.

Sure, there were exceptions to this model (the Yellow Pages and the classifieds, for example), but generally speaking, the biggest wins for a marketer happened in this arena.

We're watching it die.

The latest symptom of this death is the hand-wringing about the loss of the book-review sections from major newspapers. Book publicists love these, because it's a way of putting your book in front of people who weren't looking for it. Oprah is a superstar because she has the power (the right? the expectation?) of regularly putting new ideas in front of people who weren't looking for that particular thing.

Super Bowl ads? Another example of spending big money to reach the unreachable. This is almost irresistible to marketers.

Notice the **almost**.

In the last few years, this model is being replaced. Call it permission if you want, or turning the world into the Yellow Pages. The Web is astonishingly bad at reaching the unreachable: Years ago, the home page banner at Yahoo was the hottest property on the Web. That's because lazy marketers could buy it and reach everyone.

Thanks to the Long Tail and to competition and to a billion Web sites and to busy schedules and selfish consumers, the unreachable are now ***truly*** unreachable.

If I want a book review, I'll go read one. If I want to learn about turntables, I'll go do that. Mass is still seductive, but mass is now so expensive that marketers balk at buying it. (Notice how thin *Time* magazine is these days? Nothing compared to *Gourmet*.)

And yet. And yet marketers still start every meeting and every memo with ideas about how to reach the unreachable. It's not in our nature to do what actually works: Start making products, services, and stories that appeal to the reachable. Then do your best to build that group ever larger. Not by yelling at them, but by serving them.

Interruption = Spam

Consumers have more choices than ever before. More media choices, more choices of products and services. There aren't three TV networks; instead, there are a million (literally) things to watch on YouTube. There aren't a dozen radio stations; there are a million (literally) online. As a result, the consumer has the power to say, "If I'm not interested in what you have to say, I won't watch it. I'm not a hostage any longer."

E-mail spammers did consumers one great service: They taught people that it was OK to hit Delete. Just as the VCR taught people that it was OK to fast-forward past commercials, spammers taught consumers that it was OK to forcibly skip all the ads, and things would be just fine.

So people do. Smart consumers now skip all the ads.

And yet almost all marketers continue to spend on interruption media. We do it because it's traditional and safe and easy and what the boss and the client expect. We rarely do it because it works.

At the same time that many organizations waste billions of dollars interrupting strangers, a few companies are busy building a permission asset—the privilege of delivering anticipated, personal, and relevant ads to the people who want to get them.

Outbound marketing now demands respect for the people on the receiving end.

Permission Is Not a Marketing Tactic

The organizations that attempted to use permission marketing to augment their interruptive techniques have often complained that they don't like it. It takes too much patience. It's not fast enough. They want to throttle up and make the numbers scale.

Organizations that have realized that permission is not a tactic but an asset, though, have discovered a very different result.

Pets.com used every tactic they could find—TV ads, sock puppets, and accelerated e-mail marketing with a veil of permission. They blew through a hundred million dollars by interrupting people who didn't want to hear from them. They failed.

Dailycandy.com did the opposite. This daily e-mail newsletter has hundreds of thousands of subscribers to their fresh e-mail newsletters on style. They are extremely profitable and growing every day. They used no TV, no puppets, and a very calm and patient approach to permission marketing. They won.

Permission Is Not Up to You

Before we move on to the second trend, I need to underscore some critical ideas about permission marketing, because they are often misunderstood. Here they are from the consumer's point of view:

1. Permission doesn't exist to help you (the marketer). It exists to help **me**. The moment the messages you send me cease to be anticipated, personal, and relevant, then you cease to exist in my world.
2. My permission can't be bought or sold. It's nontransferable.
3. I don't care about you. Not really. I care about me. If your message has something to do with my life, then perhaps I'll notice, but in general, don't expect much.
4. Privacy policies and fine print are meaningless to me. When I give you my permission to follow up, we're making a deal and you're making a promise. An overt

and clear promise. If you break that promise, whether or not you are legally in the right, we're finished.

5. I demand your respect. I can get respect from plenty of organizations, so if you disrespect me (by mistreating me, by breaking your promise, by cheating or lying, or by undervaluing our relationship), then sure, that's right, you're history.

TREND 2

AMPLIFICATION OF THE VOICE OF THE CONSUMER AND INDEPENDENT AUTHORITIES

In a market where everyone is a critic, the need to create products that appeal to and satisfy critics becomes urgent. The same is true for after-purchase issues of service and quality.

Why Have You Heard This Story Before?

Do a Google search on "Dell Hell" and you'll find hundreds of thousands of matches. You'll find entire sites devoted to discussions about Dell and their service policies. And most of all, you'll find references to Jeff Jarvis, a blogger and former magazine executive.

It's not just Google. Pick up any number of books on new media and there's Jeff's story again. A story about Dell ignoring his pleas, cries for help, and eventually anger, until finally they realized what a hit they had taken. A billion-dollar brand, besmirched by a guy in his pajamas.

Why such a fuss over a three-thousand-dollar laptop?

Jeff's Dell is a symbol. Like the storming of the Bastille, it has become a totem for individuals who want to remind the Fortune 500 (and beyond) of consumers' increasing power. It's not just that individuals are discovering how much power their amplified word of mouth has. It's that they have wanted it for so long that they are giddy about it.

For a hundred years, mass marketers have ignored individual consumers. These companies realized that they have the money, the marketing dollars, the distribution channels, and the factories. A few malcontents are really no big deal. For a hundred years, "That's our policy" was generally considered an acceptable response to a disgruntled customer.

Consider the 1 Percent

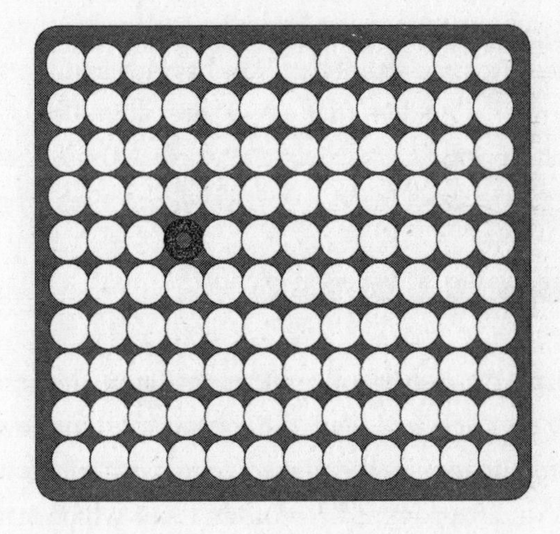

Ben McConnell and Jackie Huba's book *Citizen Marketer* proves that in just about every community, 1 percent of the people are the givers.

In Wikipedia, for example, about 1 percent of the users create and edit articles. Same goes for Microsoft's Channel 9 Web site. They get four and a half million visitors a month, and almost exactly 1 percent of them contribute comments. The same math works for Digg, Reddit, and YouTube. One percent of blog readers are blog writers. One percent of talk-radio listeners are callers.

The thing is, you don't know who they are. You don't know which 1 percent of your customers and prospects are the ones who need to, love to, and want to post about their experiences.

It's like Russian Roulette. You have to assume that every chamber is loaded, that every interaction is an interaction with a critic.

Joanne Is Coming!

One of my dearest friends is Joanne Kates, the restaurant critic for the *Globe and Mail*, the most important newspaper in Toronto. Joanne carries a credit card with someone else's name on it. (I promised I wouldn't say which name she uses.) Despite her precautions, her picture is posted in the kitchens of dozens of top restaurants. Why? Because once a chef knows that Joanne is wearing a wig and sitting in the dining room, the staff can influence the review.

Once a server knows it's her, he can make sure the service is perfect, the food is hot, and the check is calculated properly. Once he knows it's her, he can guarantee that the staff will do their best.

You've already guessed the problem with this strategy. The problem is Zagats (and Chowhound.com, and a thousand other restaurant blogs). There isn't just one Joanne Kates in Toronto anymore. Now there are thousands.

You can no longer be on the lookout for Joanne. Now you have to be on the lookout for everyone.

Is This on the Record?

Out of the 300 million people living in the United States, just a handful are professional journalists. If you were a company or a politician, there were just a few thousand reviewers or columnists or TV personalities you had to worry about.

That meant that, most of the time, mediocre service went unnoticed. Sure, someone might tell a few friends, but that was it.

No longer.

Today, when a Comcast installer falls asleep on a couch during a job, it's likely to be videotaped and uploaded to YouTube. Today, when Procter & Gamble launches Crest Pro Health mouthwash, P&G doesn't just have to worry about word of mouth among dentists. That was manageable. Consumer feedback is a different story: The product has already received more than forty reviews on Amazon, almost all of them startlingly negative (it turns your teeth brown, apparently). Here is one of the more **positive** reviews:

> It can kill your sense of taste (foods taste kind of "metallic" the next day after you use it). . . . It does stain teeth. I've had perfectly bright and shiny teeth all of my life. After using this product, I started noticing small brown stains appearing on several teeth. Fortunately they are more towards the middle and back, but I am still not very happy, and I hope these stains fade with time.

With friends like these, it's not clear you need reviewers.

The fact is, you're always on the record, everyone is a critic (or could be), and the Web remembers forever.

Why Blogs Matter

You're forgiven if you don't get it.

You don't get it because, like me, you grew up with Ultraman and Batman and That Girl. You grew up with Dan Rather and Walter Cronkite. You grew up with "and we'll be right back after this word from our sponsors."

We were readers.

They were writers.

You read novels written by strangers. Saw movies directed by strangers. Believed the news delivered by strangers.

Those guys were professionals. They were middle-aged, white, hirsute (where it counted), and well paid. Paid for by the commercials that we believed.

So blogs are like CB radio. Blogs are this hobbyist side-show, and most of them, while amusing, aren't particularly relevant.

It's easy to write the whole thing off or, even worse, to try to manipulate it by sending out free laptops to influential bloggers.

I think it's a lot bigger than that.

Here's the fundamental shift that I hope every marketer will understand: For the first time ever, blogs convert readers and viewers into writers. And YouTube turns them into directors.

That change in posture undermines the fundamental core of almost all traditional marketing theory. It's not us and them. Now it's us and us.

Oof.

Here's what to do if you still don't get it: Start a blog.

Fine with me if you want to do it under an assumed name. But do it. About something you care about. Then, check back in with yourself a month later and see if you're different. And how you're different.

Understanding Blogs

There are more than 80 million blogs worldwide today tracked by technorati.com. You'd think that people would have figured out what a blog is by now.

Because it's a tool, not a thing, it's easy to get confused. Sort of like, "what's a computer?"

For most bloggers, a blog is just one thing: a personal publishing platform. A place where one person gets to talk.

What makes blogs a lot more interesting than the many other personal publishing platforms that came before (Christmas letters . . . writing to the local newspaper . . . Geocities . . .) is that blogs are connected, tracked, indexed, and spread around. One article in one blog might be surfaced by another blog, which leads to exposure to a dozen or more readerships. Suddenly, and forever, that article is singled out, digested, and influential.

Let me be superclear here: A post on a blog anywhere in the world could very well rank higher in a Google search than information on that same topic on your company's Web site. Which means your point of view disappears, and the point of view of some blogger comes across instead.

Most blog posts are innocuous personal items of little interest to anyone but the poster. But the platform doesn't care. It supports those posts just as well as it supports the detailed review of your call center.

When Asi Sharabi did a survey of the one hundred most-viewed videos on YouTube, he discovered that 58 percent of them were created by users, not repurposed TV shows. Even on video, people want to hear themselves talk. We want to hear what other people have to say.

The easiest way to understand blogs (text, audio, or video) is to understand that they (finally) connect three real desires: to hear our own voices, to be heard by others, and to hear what the crowd thinks.

For the first time in history, the graffiti on the wall has more power than the official version coming from an organization (even yours).

Should You Have a Blog?

It depends.

If you have something interesting to say and the guts to say it.

If the people you want to talk to have significant interest in what you're saying and the time and access to your blog to read it.

If you are willing to tell the truth.

If you can stick with it for months, or more probably, years.

The shift is obvious: ads are about you, a blog is about your readers. Ads are about control from the top, blogs are about honest communication from one person to an interested, sometimes passionate, always opinionated audience.

Getting Dugg

Digg.com isn't like anything.

I mean, Amazon is like Barnes & Noble, just online. And Google is a little like a really amazing library. But Digg doesn't have a real-world equivalent.

Digg gets 8 million unique visitors every month. Once users get to the site, they take a look at the work of the 80,000 citizen editors who have volunteered their time to pick the coolest, most timely posts from across the Web.

In a typical day, you might find the following:

- Rudy Giuliani in Drag Smooching Donald Trump
- Full Nintendo Wii Manual Scanned & Uploaded
- *Animator* vs. *Animation II*

- Retailers See Strong Mac Demand, Little Excitement over Zune
- High Resolution Cropped Halo 3 EGM Scans—Great Quality
- Uranium and Plutonium Traces Found in Iranian Facility
- Mario (Charles Martinet) Does Movie Impressions at E3
- The Colbert Report: Interview with Dan Rather
- $12B I.T. Project "Sleepwalking Toward Disaster"
- Baghdad: Gunmen Dressed as Police Commandos Kidnapped up to 150 Staff . . .
- America's Closest Allies Now Believe Bush Is a Threat to World Peace
- EFF Asks Supreme Court to Tackle Secret Law
- How to: Make a DVD Burner into a High-Powered Laser

No individual could ever come up with this list. But eighty thousand competitive geeks with too much time on their hands did. And millions of people saw the list and responded to it.

Showing up on the front page of Digg means that you've received more votes than any of the other thousands of pages nominated at the same time. This also gets you about twenty thousand new visitors to your site in less than a day.

The good news is that this group of twenty thousand people is educated, motivated, and in a hurry. The bad news is that these visitors don't necessarily buy from you, click around your site, or respond to ads.

At Digg's current rate of growth, they'll soon have 100 million users. Hundreds of thousands of people interested in your idea. Your job will be to convert that momentary attention into long-term permission and then into action.

The question, then, isn't how you get Dugg. The question is, how do you make stuff worth Digging?

If your organization has ever issued a press release, you need to rethink that strategy in light of Digg and the other social news sites online.

A press release is designed to get one editor to assign your story to one reporter. That reporter is supposed to write a story about you that gets read by hundreds of thousands of people. A few of them might care enough to buy your product.

Now, though, you're not issuing a press release; you're alerting a small group of consumers who have given you permission. They are your blog readers or RSS subscribers. They individually make a decision that ends up being seen as a group action—they Digg you! And if the story is good, the next level of Digger sees it—these are total strangers, motivated to action by the power of your news. And on and on until eventually twenty thousand or fifty thousand or more people have visited your site to see what it's all about.

Digg does more than put numbers on a phenomenon that's been around forever (word of mouth). It dramatically amplifies it.

Digging Brian's Copy

Brian Clark runs a Web site called Copyblogger.com. On it, he runs ads to pay the bills, but mostly his goal is to "meet cool people with great skills and great ideas and do business with them." It's working. His site has grown to more than six thousand subscribers in less than a year, mostly by practicing what he preaches.

Some readers will dismiss six thousand subscribers as a tiny number. It's not. Most trade magazines have fewer than twenty thousand subscribers, and most of those subscribers don't even bother to open the magazine. Brian, on the other hand, has a direct, emotional, regular connection with a hundred times as many potential clients as he needs to be fully employed for the rest of his life. Big numbers matter when you run mass ads that are ignored. Small numbers are just fine when you deliver powerful messages with permission.

Brian writes diggable articles. His readers (who have given him permission by subscribing) often go to Digg and post them. This brings tens of thousands of new readers, some of whom subscribe.

Over time, readers (like the subscribers to Brian's blog) tend to trust the people who write for them. Doesn't matter if it is a magazine, a series of novels, or a blog. Trust comes from repeatedly delivering insight and truth. Brian tirelessly interacts with his readers, and eventually those interactions will turn into projects and profits.

Flipping the Funnel

Every business has its 1 percent. Every business has a group of customers so motivated, so satisfied, and so connected that they want to tell the rest of the world about you and what you do.

Your challenge is to give these people a megaphone. To switch your view of the market from a vertical funnel (attention in at the top, sales out at the bottom) to a horizontal one, in which ideas spread from one prospect to another.

The Incredible Power of Synchronicity

People have always wanted to do what other people are doing. We feel safe and secure and validated when we choose the popular records, vacation spot, or clothing. Sure, some people choose to stand out, but most people, most of the time, want to be like most people, most of the time.

The Web has surfaced all of these trends for easy inspection.

I can see which blogs are popular, which Amazon items are selling, and which candidates have raised the most money. I can do it every day, and I can do it instantly.

Not just across the mass market (that's what *Variety* reports in their box-office figures) but in my microcommunity. I can easily see which Thai cookbook is the bestseller. I can discover which Squidoo lens has the best ratings and which one seems to be ignored. Within my world, whatever **world** that is, I can see who is winning.

Anyone who has been through basic training has learned how to run across a bridge. You must change your cadence so that everyone doesn't step at the same time . . . the synchronized footfalls can easily destroy the bridge. Online, we're seeing voices synchronized and thus amplified. Each voice is a little louder, a little more important. That means the rest of us can see what's hot.

And we make our decisions accordingly.

TREND 3

NEED FOR AN AUTHENTIC STORY AS THE NUMBER OF SOURCES INCREASES

Consumers hear about organizations from many sources, not just one. As a result, you have to get your story straight. Saying one thing and doing another fails, because you'll get caught.

Why George Allen Won't Be Running for President

It was a great Web moment. George Allen was the Republican Party's next star, anointed as a potential candidate for president in 2008. But first he had to win the Senate race in Virginia, considered by many to be a layup for him.

The traditional way to run a political campaign is to control your message. Control what you say and when you say it. Control who hears it.

Tell one story to your raving fans, and a more moderate story to people in the center.

As voters have seen again and again, politicians are good at this. Some people call it lying. But in general, politicians have gotten away with it.

The top-down, control-the-message strategy worked in the past for a few reasons:

- Media companies were complicit in not embarrassing the people they counted on to appear on their shows and authorize their licenses.
- Politicians could decide where and when to show up and could choose whether or not they wanted to engage.
- Bad news didn't spread far unless it was exceptionally juicy.

But George Allen discovered that the rules have fundamentally changed. Allen's challenger asked S. R. Sidarth, a senior at the University of Virginia, to trail Allen with a video camera. The idea was to document Allen's travels and speeches. During a speech in Breaks, Virginia, Allen turned to Sidarth and said, "Let's give a welcome to Macaca, here. Welcome to America and the real world of Virginia," said Allen. As I write this, YouTube reports that Allen's slur has been watched on YouTube more than 318,000 times. Add to that the pickup from the broadcast media (which picked it up because it was popular, not because it was "important"), and you see why George Allen lost the election.

The first words out of Allen's mouth on the tape are, "Ladies and gentlemen, we're going to run a positive campaign." The story didn't match the facts, and the facts showed up on YouTube.

Telling Two Stories at Wal-Mart

Wal-Mart is the most profitable company in the history of the world. They've done a great job of increasing the standard of living of their customers by lowering the prices of what they sell. At the same time, they've been spending millions of dollars in marketing the idea that they take excellent care of their workers. In the words of one Wal-Mart executive, "Health care is the most pressing reputation issue facing Wal-Mart." By running ads featuring the stories of happy workers, Wal-Mart thought they could spread the word and improve their reputation.

That story, with plenty of repetition, would eventually sink in. Americans, usually a complacent and greedy group, would take the bargains, smile at the clerks, and drive home. Until an internal memo surfaced.

M. Susan Chambers, Wal-Mart's executive vice president for benefits, wrote,

> Growth in benefits costs is unacceptable (15 percent per year) and driven by fundamental and persistent root causes (e.g. aging workforce, increasing average tenure). . . . Unabated, benefits costs could consume an incremental 12 percent of our total profits in 2011, equal to $30 billion to $35 billion in market capitalization.
>
> While Associates are satisfied overall with their benefits, they are opposed to most traditional cost-control levers (e.g. higher deductibles for health insurance). Satisfaction also varies significantly by benefit and by segment of Associates. Most troubling, the least healthy, least productive Associates are more satisfied with their benefits than other segments and are interested in longer careers with Wal-Mart.

The memo encouraged Wal-Mart to simultaneously increase employee satisfaction and decrease costs by encouraging turnover and hiring new workers. The new employees, it turns out, are a lot cheaper than the old ones and need less in the way of benefits. Quick! Fire the old and the sick.

The memo surfaced in the *New York Times* and then raced through the Internet. (It shows up more than twenty thousand times on Google.) Suddenly, telling two stories at the same time is getting very difficult.

But What Does Your Broker Think?

The Wall Street analyst system was a golden goose. Every day, high-priced research analysts turned out stock reports based on inside information. These mandarins had access to CFOs and other company officials, as well as the time and skill necessary to do a lot of number crunching.

As a result, the best customers at the big firms (like Merrill Lynch) got access to great data. Using that information, they could make smart trades and earn more money. Everyone came out ahead—the investors (who made a profit), the brokers (who made a commission), the analysts (who made a fortune), and the investment bankers, who used the entire system to lure companies into giving them lucrative projects.

Then, the other half of the story came to light. Basically, companies were buying positive recommendations in exchange for steering business to one firm or another. The reports were rigged.

In the fallout that followed, the entire system was denatured. As soon as the cover story turned out to be false, no one wanted to risk being fooled again. Investors fled to discount brokerage houses, analysts lost their jobs, and the transparent economy again proved its power.

It's Not Just About Scandal

The three examples I've chosen so far are the big ones, the juicy kind that get headlines. More often than not, though, the disconnect between story and reality happens far more subtly. The only people who notice are potential new customers.

It could be something sensational, as when Emily Gillette was thrown off of a Delta Airlines flight for breastfeeding. Word spread worldwide, and Delta can't possibly run enough ads to make up for it. Or it could be something much sillier, like the video of the sleeping Comcast Cable installer that I mentioned earlier—it has been seen more than eight hundred thousand times (http://www.youtube.com/watch?v=CvVp7b5gzqU).

More likely than not, though, it's the stuff you've never

heard of. It's the really nasty UPS driver who manages to turn off dozens of potential customers every day as he does his route. Or the local restaurant with the offensively worded menu that prevents local moms from bringing their kids in for sushi.

As consumers, we're getting better and better at discerning the difference between the fake and the real. We look for a can opener in the kitchen of the fancy restaurant, or the telltale return address in an e-mail that appears to be from a real person.

Sure, some people embrace the faux. We stand in line to eat at an Applebee's near the mall. But when consumers are making a decision that matters to them, they often rely on the truth as they receive it from the community, not on the story the marketer manufactured.

The Internet Doesn't Forget

Landlords have learned (the hard way) that renting to a bad tenant is an expensive mistake. So they are using the Web to learn about people before they sign a lease.

On-site.com runs fifty thousand background checks every year for landlords in Manhattan alone. Forty percent of the applicants are either turned down or given a rank of Maybe.

The Web knows, and it doesn't forget. It knows how much money people make, whether they've been to housing court, and how long they were in their last job.

The Web remembers what an unhappy customer posted about your customer service people—even if that event took place three years ago. Sort of like Santa Claus, but without the gifts.

This shared institutional memory is something that has never occurred before. I can quickly see the reputation of each person at a company I'm dealing with, just as easily as I can see what a celebrity looked like before his facelift.

Stories Spread, Not Facts

People just aren't that good at remembering facts. When people do remember facts, it's almost always in context.

Patagonia makes warm coats. So do many other companies, almost all of which sell their coats for less money, do less volume, and turn a lower profit.

Is it because Patagonia coats are more beautiful or warmer?

Not at all.

It's because the company has created (and lives) a story that has absolutely nothing to do with clothing and everything to do with the environment. The company has gone all the way to the edge, further than just about any other major brand, in building an organization that reduces, reuses, and recycles. A company that gives back (in terms of attention and cash) to the environment at the same time that it challenges it. The catalog is filled with beautifully illustrated stories of some of their climbers—these guys really are nuts.

As their clothes have become less flashy and more organic, Patagonia has grown. An organic cotton shirt from Patagonia costs ten times more than a similar nonorganic, nonbranded shirt from someone else. And that's fine. Because the shirt isn't just a shirt—it's a story and a symbol. It's a way for a wealthy consumer to tell herself a story about her priorities, at the same time she tells that story to her friends.

No one talks about The North Face. Plenty of people talk about Patagonia. That's because the company, from the top down, lives and believes their story (a unique story and a story that resonates with their customers) and makes it easy for the story to spread.

TREND 4

EXTREMELY SHORT ATTENTION SPANS DUE TO CLUTTER

The death of mass marketing is partly due to the plethora of choices and the deluge of interruptions. As a result, complex messages rarely get through. (This section is extra short, of course.)

Short Books, Short Shows, Short Commercials, Short Ideas

In 1960, the typical stay for a book on the *New York Times* bestseller list was twenty-two weeks. In 2006, it was **two.** Forty years ago, it was typical for three novels a year to reach number one. Last year, it was twenty-three.

Advise and Consent won the Pulitzer Prize in 1960. It's 640 pages long. *On Bullshit* was a bestseller in 2005; it's 68 pages long.

Commercials used to be a minute long, sometimes two. Then someone came up with the brilliant idea of running two per minute, then four. Now there are radio ads that are less than three seconds long.

It's not an accident that things are moving faster and getting smaller. There's just too much to choose from. With a million or more books available at a click, why should I invest the time to read all 640 pages of *Advise and Consent* when I can get the idea after 50 pages?

Audible.com offers more than thirty thousand titles. If an audiobook isn't spectacular, minute to minute, it's easier to ditch it and get another one than it is to slog through it. After all, it's just bits on my iPod.

Of course, this phenomenon isn't limited to intellectual property. Craigslist.org is a free classified-ad listing service. A glance at their San Francisco listings shows more than 33,000 ads for housing. That means that if an apartment doesn't sound perfect after just a sentence or two, it's easy to glance down at the next ad.

A Ticket to the Game Is Cheaper Than Ever . . .

It's easier than ever to sell something. It's easier to publish a book; it's easier to market a board game. If you want to be in the soda-pop business, all you need is a sticky label and some cash. Someone else will do the bottling and distribution for you. In just about every industry, organizations have learned how easy it is to generate a profit by providing the back office for entrepreneurs who want to enter the market.

It's likely that you've never heard of Hon Hai Precision Industries. You may never have heard of their American brand name, Foxconn. Unless, of course, you work for any of the brands of home computers or cell phones. If you want to roll out 10 million phones to the U.S. market, just call Hon Hai. They'll make the phones to your specifications, and the phones they make will actually work.

You may not have heard of lulu.com, either. This startup will take your manuscript and turn it into a book—a book that looks just like one that Random House would publish. They'll give you a storefront online to sell it from, and an ISBN so you can take orders from any bookstore in the world.

Zazzle.com makes it easy for any organization to sell just about any item imprinted with their name on it. You can create an "I Love My Church" thong or a "Maxwell Studios" baseball cap. Zazzle makes them, boxes them, and ships them. You sell them on a site that Zazzle builds and you get to keep the money.

Costco does the same thing with vodka. They are selling millions of dollars of their Grey Goose–style vodka in their stores. Costco uses their house brand on the label but you can be sure they're not busy distilling potatoes.

. . . And So Is the Advertising

The explosion of media alternatives is so pervasive that it's easy to forget what an impact it has had on viewers and advertisers. You can watch (to pick a bunch randomly): Sky One, Virgin 1, E4, More 4, Living, UKTV Food, Paramount Comedy, Hallmark, The History Channel, Boomerang, Cartoon Network, Discovery Kids, Disney Channel, Toon Disney, MTV, Sky Sports, VH1, Discovery Home, UKTV History, BBC Three, BBC Four, BBC Parliament, Zee TV or a channel dedicated to your favourite football team

With all that choice, surely there's room for more advertisers, right? Glad you asked.

A company like thoughtequity.com will gladly license you a professionally shot, very expensive TV commercial that you can use as if you made it. Car dealerships, real estate brokers, and other local businesses can put together a cheap, fast ad campaign in less than a day.

It's possible to buy carefully targeted TV advertising for a few hundred dollars. It's easy to buy extremely targeted online advertising for ten cents a click.

Translation: You can buy tiny slices of attention for a fraction of what it cost just a decade ago.

The YouTube Generation

YouTube now carries more than 7 million videos. Though the average length of a video on the site is about five minutes, most videos are watched for less than ten seconds. Ten seconds is all you get to prove whether the viewer ought to invest another ten seconds, and if you get someone to stick until the end, you've hit a home run. With 7 million to choose from, there's not a lot of reason to persist through something that's not engaging.

This rapid-fire approach to media is important even if you're not a media company. It means that the personality of your product on the shelf matters. It means that every interaction with a consumer is a make-or-break proposition. You don't get the chance for a learning curve. You don't get the opportunity to have the user overcome initial discomfort.

TREND 5

THE LONG TAIL

Chris Anderson's work demonstrates that in almost every single market, "other" is the leading brand. Domination by hit products is fading, consumers reward providers that offer the most choices, and the economics of creating and selling a product have fundamentally changed.

Brand Managers Can't All Be Stupid

In my corporate speeches, I often tell the audience, "Raise your hand if your organization offers more SKUs (stock-keeping units) than it did five years ago." Every hand goes up. The point of the exercise is the sad news that more variety equals more clutter, and yet marketers are dealing with clutter by making more of it.

But are these marketers stupid? Why does every member of the audience raise a hand?

Because they're smart. Because every time they create new varieties, new sizes, new price points, and new niche products, sales go up.

This is confounding. After all, if the mass market is about mass and price, you would expect that economies of scale would lead to just a few items being made, each perfected in form, content, and price. You'd expect that there would be two or three breakfast cereals dominating the market, not a hundred.

It turns out that over time, the market doesn't settle on one or two leaders. Given the freedom, market leadership spreads out, just like an oozing cube of melting caramel. While there may still be a number one in a market, the *market* itself is a lot smaller. The number-one beer among fraternity brothers might be very different from the number-one beer among stockbrokers. Any market of people with sufficient resources will get very picky on you.

Starbucks offers nineteen thousand varieties of beverages. Why? Not because Starbucks likes the extra work or because management likes all those varieties. It's because their customers do. You can choose from 3.5 million songs on iTunes. More than you could listen to in several lifetimes. The reason is simple: In the long run, more variety always leads to more sales.

The Long Tail, a brilliant term coined by Chris Anderson in his book and blog of the same name, is a simple idea that starts with this: Given the choice, people want the choice. In any normal marketplace, if you give people more choices, revenue goes up.

And Now People Have the Choice

By itself, a bias for choice is interesting but not particularly surprising. What's surprising is the magnitude of this desire. My favorite example is the comparison of a typical Barnes & Noble store with Amazon. If you examine the sales of the 150,000 titles in a big store, you'll see that they account for perhaps half of Amazon's book sales. In other words, if you aggregate the millions of poorly selling titles on Amazon, they add up to the total sales of all the bestselling books in the physical world put together.

Another way of looking at it: More people watched more video on YouTube last week than watched the top ten shows on network television.

Another way: A quick look at your grocer's beverage aisle will prove to you that Coca-Cola is no longer the most popular soft drink in the country. The most popular soft drink is "other": none of the above. The mass of choices defeats the biggest hit.

This curve shows up over and over. It describes travel habits, DVD rentals, and book sales. Give people a choice and the tail always gets longer. Always.

The Long Tail has been around forever, but only now does it really matter. That's because of several trends working together:

a. Online shopping gives the retailer the ability to carry a hundred times the inventory of a typical retail store.

b. Google means that a user can find something if it's out there.

c. Permission marketing gives sellers the freedom to find products for their customers, instead of the other way around.

d. Digital products are easy to store and easy to customize.

e. Digital technology makes it easy to customize nondigital goods.

More Is Usually Better

John Gourville, at Harvard, and Barry Schwartz (author of *The Paradox of Choice*) have each argued that too much choice is a bad thing, that it leads to dissatisfaction and causes people to put off decision making. Too much choice makes the statement "I'll decide later because there's just too much risk of screwing up" more likely.

A trip to the mall certainly demonstrates that this is happening. Tons of empty-handed shoppers are all walking around looking for the perfect item. And often buying nothing.

It's hard to fix this problem unilaterally. You might cut *your* selection, but your competitors will just go farther out the tail.

A few marketers manage to succeed by radically limiting choice. In-N-Out Burger (which succeeds with a tiny menu at their burger stands) is one example. Other examples are companies with a unique product, like the Toyota Prius or Steinway, and a significant barrier to their competitors. But that list is awfully short. Expanding selection increases sales. When Just Born (the folks who make Peeps, those bizarre yellow marshmallow chicks every Easter) expanded their line to include ghosts, pumpkins, and hearts, sales expanded as well.

The Short Head

The familiar element of the long tail is the right part of the graph opposite, the part that goes off, slowly, forever. However, in order to have a long tail, you need a short head.

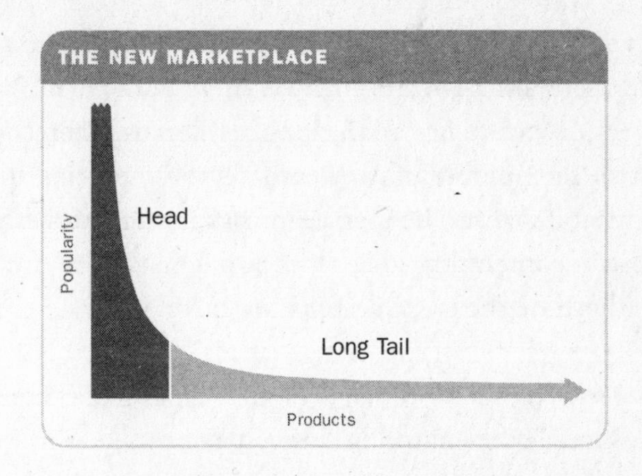

That's the part of the graph on the left, the part where the top 40 and the number-one hits live.

Yes, down along the tail there are fifty little minihits that equal that big hit, but that big hit still matters. Bestsellers fuel the long tail wherever it lives. Whenever Netflix rents three copies of *The Gabby Hayes Show: Vol. 2 (1954)*, they also end up renting three thousand copies of *Crash*.

You may decide that you want your organization to be in the short-head business. One apocryphal study done by McKinsey for a major book publisher recommended that the company publish only bestsellers. Of course, this is absurd. If you could just publish bestsellers, then you would. But you can't.

In my previous book, *The Dip*, I argue that in every single micromarket, the market rewards the leader, the best in the world. Winners still win, and they win more often because there are far more markets in which to win. Your challenge is not to ignore the long tail. It is to embrace the fact that the number of markets is skyrocketing and to invest what it takes to be seen as the best in any market you choose to compete in.

So here are the two questions we're left with:

- How big a percentage of our time and resources should be spent trying to invest in creating short-head blockbusters?
- How can we reorganize our efforts to better support the long tail?

If both the hits (the short head) and the long tail are equal contributors to your business, why focus all your effort on the former?

Once again we see that critical marketing decisions are being made by people who have nothing at all to do with the marketing department.

The Original Long-Tail Retailer

In 1897, Sears Roebuck offered a catalog with more than 300 pages of goods; it ballooned to more than 1,400 pages in 1975. This was a mail-order company that sold watches, underwear, and yes, houses. In fact, more than 100,000 houses were sold by mail order before the catalog and the houses were discontinued.

No fancy technology, no triumph of consumer interaction, the Sears catalog merely represented a fundamental desire—to have exactly what you want, when you want it.

All the Internet has done is make the fulfillment of this wish universal. Many consumers spoiled by Amazon have a very hard time shopping at a "real" store. They know they ought to be able to instantly search the entire inventory, to find relevant items in proximity to the ones they are already looking at, and to have every single item available for purchase at the same time in the same place. They also want to buy it all with one click.

What Richard Sears discovered is that a catalog frees a retailer from physical constraints. By putting a retail store in millions of homes (and having a centralized warehouse), you can profitably stock far more items than a traditional store ever could. The Internet takes that equation and amplifies it. It permits niche sellers to pursue the same strategy that Sears took, but even deeper.

The Rejuvenation Story

In 1977, a junk store in Oregon decided to try selling period lighting (the illumination of choice while reading your Sears catalog in 1910). Since then, Rejuvenation has grown to have more than two hundred employees and become the largest manufacturer and seller of period lighting and architectural fixtures in the world.

If, for example, you need a twelve-inch copper fabric bridge lampshade, with an UNO fitter (they also sell the clamp-on variety), odds are this is the only place you will find one. And if this is what you really need, then you'll find it.

And those are the two secrets of being a long-tail provider. People who really care will find you, and going to the trouble to make what someone wants is in itself a useful marketing exercise. It's not about what ***you*** think the market wants, or what ***you*** want the market to want. It's about creating and assembling a collection of goods and services that captures the attention (and commerce) of the people who truly care.

Finding Silos

A key tenet of marketing is that it's more profitable to make and sell something that large groups of people want. If there's a significant market demand for silver dishwashers instead of polka-dotted ones, make the silver ones.

Marketers have known about biggest silos for a long time. Lots of people want to buy roses, life insurance, snack foods, or fancy hotel rooms. These are so common that these markets are obvious and lots of organizations have focused on serving this demand.

What the Web is turning up are interests that we didn't expect or plan on.

Here, in alphabetical order, are the top ten searches on CafePress:

Anti-Bush
Army Wife
Autism
Costume Alternatives
Dance
Kokopelli
Marines

Nuclear
Vegetarian
Weenies

There are hundreds of thousands of CafePress searches, but these show up often enough to make the top ten. Given the choice of just about everything, this is what people choose. Notice what's not here: big Hollywood stars or sports personalities; the usual hits are missing.

No, I don't think the Kokopelli items are being created by people who are sure there's a big market out there for them; they're just people who like Kokopellis. But it turns out that there *is* a big market for them. A huge one, in fact. Who knew? The Web knew.

As media continue to fragment, the giant obsessions (Jennifer Aniston) don't always dwarf the little ones. The Web allows other, minor obsessions to coalesce. New markets emerge, markets that are terrifically profitable but were previously unknown.

The three challenges are straightforward:

1. Find a market that hasn't been found yet.
2. Create something so remarkable that people in that market are compelled to find you.
3. String together enough of these markets so you can make them into a business.

TREND 6

OUTSOURCING

It's not just **possible** to find someone to make/code/do something for you quickly and cheaply; it is now easy. The means of production of physical goods and intellectual property is no longer based on geography but is based on talent and efficiency instead.

When Did You Last Buy from Flextronics?

It depends on whether you have a device from Microsoft, HP, or dozens of other companies. Flextronics makes products for many huge brands. Not just chips but entire products. And why not? They produce high-quality products, to spec, cheap.

When you buy an article of clothing, an appliance, tax preparation services, or a computer, chances are you're not buying from the person you think.

There are excellent economic reasons for this. Comparative advantage means that a company organized to do nothing but produce from plans—and to do it in a low-wage, high-skills location—will always outperform a vertically integrated marketer.

Just as important, though, is the idea that a marketer without a factory is actually more innovative, faster moving, and more fashion-focused. Why? Instead of trying to keep the factory busy, the marketer can focus on keeping the market busy instead.

Mohit Gupta for Hire

Banwan, a student from South Africa, posted his assignment to write an essay about the English Civil War on an outsourcing bulletin board. Mohit, working from India, accepted the job, and for sixty pounds, he did Banwan's homework. Mohit is quoted as saying, "I worked for three hours to accumulate the history. I had to get the material in the required format and detail the causes, consequences, and the important events of the war." It might not be ethical, but it's easy.

Outsourcing the Search for Votes

Josh Hallett reports that Katherine Harris (she of the Gore recount) hired people in India to visit blogs on her behalf. They posted comments anywhere they saw a post that was negative about Katherine or her campaign for Senate. Stuff like, "Katherine Harris is tested by fire. She is strong and determined and will not back down in the face of adversity."

It didn't work, but you can bet it was cheap.

Just because you *can* outsource something doesn't always mean you *should*. The ease of outsourcing puts pressure on traditional suppliers to do something more than be convenient. As Katherine Harris learned, building a grassroots movement isn't the same as making iPods.

What Katherine Harris learned as a candidate is something that every organization is going to learn soon: You can harness the power of thousands of people for very little money. But this is no substitute for honesty, passion, and authenticity.

We Can Pray It for You (Wholesale)

The *New York Times* reports that for decades, the Catholic Church has been outsourcing the duties of saying special prayers and requiems to priests in Kerala, India.

According to the *Times*, the Reverend Paul Thelakkat, a Cochin-based spokesman for the Synod of Bishops of the Syro-Malabar Church, said, "The prayer is heartfelt, and every prayer is treated as the same whether it is paid for in dollars, euros, or in rupees." It's worth noting that a prayer purchased locally costs about ninety cents, while the standard price in Europe or the United States for the same prayer is five dollars.

Not Just India

Pull into a McDonald's drive-through in high-wage Oregon, and you'll find yourself talking to the same mechanical-sounding voice you're used to. Except that the voice is in Grand Forks, North Dakota, and is being beamed across the country via the Net. You place your order, your car is photographed, and the picture and the order show up on the screen of the cashier at the next window, who is ready with your food and your bill.

By pooling the inputs of several McDonald's restaurants to one call center, they end up eliminating downtime at the same time they can do better training and quality control.

When you look at these examples, it's easy to feel unsettled. Unsettled about the ethics of doing your homework via a long-distance worker, or just unsettled about the easy jobs for teenage girls that Oregon lost to North Dakota.

The ethics questions are real, but the competitive questions are even more immediate. What happens when your industry's cost structure is reduced by 50 percent? Just as aggressive use of information technology and computerization increased productivity a decade ago, no industry is going to be untouched by the move to outsource anything that saves time or money.

The Real Challenge of Outsourcing

It's clear to me that there are only two paths. One path is to take every repetitive, by-the-book task in your organization and outsource it or mechanize it. The other path is to take every repetitive, by-the-book task in your organization and give the people who do that task the freedom, the incentive, and yes, the imperative to do something that cannot be outsourced.

Either what you're doing is repetitive, in which case you ought to outsource it, or it's homemade, insightful, and filled with initiative and judgment, in which case you can charge for it.

It's not fair. Your competitors are busy cutting costs by outsourcing the work that you struggled to make cheap in the first place. But if a job **can** be codified, it **will** be outsourced, usually for less money.

TREND 7

GOOGLE AND THE DICING OF EVERYTHING

Google and the other search engines have broken the world into little tiny bits. No one visits a Web site's home page anymore—they walk in the back door, to the page Google sent them to. By atomizing the world, Google destroys the end-to-end solution offered by most organizations, replacing it with a pick-and-choose, component-based solution.

Not to Be Sold Individually

It's not so sexy, and it doesn't get as much press as many of the other trends on this list, but it is probably the single most important idea, the one that drives the others.

Bundling was the glue that held together almost every business and organization.

Bundle donations and parcel them out to charities that deserve them.

Bundle TV shows and present them, with ads, on your TV network.

Bundle the items in your industrial supplies catalog and hand it to the business buyer.

Bundle thirty businesses and house them in one big office tower.

The Yellow Pages is a multibillion-dollar business that consists of nothing but bundled ads for local businesses. No one wants to keep a flyer for every business in town, but everyone has a copy of the Yellow Pages.

Book publishers bundle authors and share the expertise of their staff, their sales force, and their capital in order to bring books to readers.

We've been doing the bundling so long, we forgot we were doing it.

Atoms, Not Molecules

And then came the online search. Not gross searches for big ideas (like "industrial supplies") but specific, directed searches for particular needs (quarter-inch copper washer) or specific items (ISBN 12–23232–0x).

Specific searches mean that bundling is not necessary. Not only do we not need the bundles, but in most cases we don't want them, either.

The bundles slow us down in our search for precisely what we need.

The bundles cost us money, too, because we won't allow our item to subsidize something we don't want.

An interesting sidelight: Even though many people (even most people) don't find everything via Google, the group that does is big enough to have already changed the world for everyone else. Once a marketer or a manufacturer or a group realizes that they must present their atoms for inspection, the philosophy behind bundling starts to fall apart. The trade-offs we used to be able to make in order to sustain a bundled approach to the world cease to be viable in the face of all these disparate atoms, each on their own, each available to anyone who wants to grab them.

Accessories Sold Separately

I needed a car charger for my Nokia 6230 cell phone.

Back in the day (say 2004), that would have meant spending twenty-nine dollars for a new one at the local cell-phone store.

Cell-phone chargers are a bundled item. Buy a cell phone and a contract, and they throw in the charger for free. But what if you lose your charger? Well, the local store is happy to sell you one for thirty dollars.

Fast-forward a few years, and now there's no need to head over to the bundler. Instead, do a Froogle search on **car charger Nokia 6230**. What's this?—a charger for a penny! And several perfectly good ones for three dollars each. The one I found on Amazon was a whopping four dollars, but since they already had my credit-card info on file, I clicked once and boom, I'm done.

Having something in stock is no longer enough reason to charge a 600 percent markup, because now someone one click away also has it for sale. If you can't offer more than a commodity, someone else will sell it cheaper.

Profits in Broken China

Replacements, Ltd., is a business that couldn't possibly have thrived in the 1960s. Started in 1981, the company now has more than 11 million pieces of china, silverware, and knickknacks in stock. Computers make it work. Without computers, staff members would never have found anything. At first, most of their orders came by phone. "Hi, I broke a dish from my grandmother's tea service. . . . Do you have a replacement?" (They still get thousands of phone calls every day.)

Obviously, almost no one needs a gravy boat from a set of Abalone china with the Flamingo Rose pattern (it's thirty-eight dollars at Replacements). But the people who do need one will find Replacements' phone number and track it down.

eBay provides a perfect outlet for Replacements' nearly endless inventory of items and their database of photos. Instead of having the Replacements staff combing through the database, eBay allows consumers to do so. At last count, more than forty thousand people had left feedback on the results of the company's eBay auctions, meaning that Replacements has had several hundred thousand sales on eBay alone.

Replacements probably makes more money selling one plate than most china companies make on an entire set. But how to find that one plate? eBay and Google provide the lens that allows a motivated but clueless searcher to find exactly the right plate at the right time. Google has taught us that the last piece of the puzzle, the most vital piece of information, is usually worth the most. Google empowers the long tail that makes Replacements a viable (actually, extraordinarily profitable) business.

Taking It Apart

The idea that one product line can subsidize another is built into the way many businesses work. The fact that middlemen profit by bundling information or product lines or shelf space opportunities is at the heart of what makes so many businesses successful. And now, quite suddenly, a different sort of business is being rewarded. One where bundling and middleman services are deliberately undervalued.

TREND 8

INFINITE CHANNELS OF COMMUNICATION

Even with the near-total chaos most media confront, the chaos is certain to get even worse. New forms of publishing, communication, and interaction will arrive in an already cluttered world. Some organizations will thrive from this increased chaos, some will be unprepared, and some will merely fight it and lose.

What Sort of Blender Do You Own?

Here's my guess: You're not sure. It's probably a Waring, but it might be something else. You hardly notice your blender, in fact.

That would definitely not be your answer if you owned a Blendtec blender. The Blendtec is used by smoothie stands and restaurants, but what really makes it stand out are the marbles.

The folks at Blendtec have taken advantage of the myriad new channels of communication by creating a series of short videos called "Will it Blend?" Each video features Tom Dickson, CEO of the company, dressed in a white lab suit. Tom's accompanied by really cheesy 1950s music as he proceeds to blend golf balls, coke cans, a rake handle, and credit cards. In one memorable episode, he blends an entire rotisserie chicken (with some Coca-Cola to make it drinkable).

George Wright, head of marketing at Blendtec, got the site made. His webmaster posted the videos on YouTube. Total cost: perhaps six hundred dollars (including the cost of buying the video camera—and the rake and marbles!). Within five days of the launch of the site, it had spread worldwide and reached hundreds of thousands of people. Viewers to date: more than 62 million.

One of the things George built into the site was a button where viewers can suggest items to be blended in future videos. So many votes came in that he had to disable the e-mail link to his BlackBerry. (And no, they're not going to do a toupee or a set of dentures. No need to keep submitting those.)

The benefit is simple. "Nobody is questioning that we make the best blender in the world," says Wright. Once you've seen the "Will It Blend?" series, it's hard to imagine ever buying a different brand of blender.

Understanding the Zen of Venn

In order for a consumer to make a decision, two things need to happen. The second is that she needs to determine that it's worth the time or money or risk to take action. But first, she needs to know about the opportunity.

A great cruise bargain might be available, with seven nights in Rio for ninety-nine dollars, but if you don't know about it, you can't decide to go.

During the golden age of TV-driven Old Marketing, the diagram looked like this:

If you had enough money, you could get your message in front of everyone. (That's the "Notice it" part of the circle.) The challenge wasn't in reaching the masses; it was in creating a product that would *appeal* to the masses (the group that is in the "Do it" segment). You could reach the big circle with certainty, and the goal was to find products that would interest the largest number of people you were reaching. Niche products were hard to justify, because the same ad could sell a mass product to far more people. The bigger the "Do it" circle, the more successful you were.

The bigger circle represents big media. It is the audience you can pay money to reach. The inner circle represents the people who actually buy what you sell. Obviously, your goal is to make that little circle as big as possible. And the best way to do that is by making average stuff that appeals to the masses.

Even though the "Notice it" circle is a lot smaller, the "Do it" circle is bigger. That's because instead of interrupting the masses, you're getting people who care.

With the explosion of media choices, though, the diagram has changed. Now it looks like this:

You can be incredibly selective about your messaging today. You can advertise only on a blog about antique cameras, or post a podcast that only rabid fans of the local high-school football team will listen to. Instead of reaching everyone (because you have no other choice) and creating generic products for large audiences, you can now reach a tiny slice of the market—just the people who are passionately interested in your products and services. No, you won't reach everyone who might be interested in what you have to sell, but you do have the opportunity to reach the people in that zone of overlap. And you can reach them more cheaply than ever.

What AdWords Changed

Google AdWords is a very simple idea that's surprisingly little understood. On every page of Google search results, in your Gmail and your Froogle results, and more and more on the pages of other Web sites (like Squidoo or the *New York Times*), you'll find these ads.

The AdWords are smart. They appear based on the context of what you're doing. Search for "Bextra" in Google and you'll find plenty of articles about this discontinued pain reliever. But look over at the ads and you'll see that many of them belong to law firms. These firms are paying handsomely for your attention. They are filing class action lawsuits on behalf of people injured by Bextra, and the law firms figure that the very best way to reach those people is to find them at exactly the same moment that those people are looking for them. In other words, instead of racing around trying to generate attention, the firms merely stand by and wait for attention to find them.

Not since the Yellow Pages has there been a ubiquitous directory that brings together the searchers and the sought.

Not only do AdWords show up at the right time, but they are also priced intelligently. The Yellow Pages charged based on the size of the ad, and you paid regardless of whether the ads worked or not.

For AdWords, on the other hand, Google charges by the click. This means that the advertiser determines what it's worth to get a visit from an interested, qualified, and motivated consumer and pays exactly that. If someone else is willing to pay more, they get the traffic instead.

The bidding system means that the advertisers with the most motivation pay the most for top billing. At the same time, Google will adjust placement based on how many times an ad is clicked on. As a result, the ads that run the most are focused, relevant, and beneficial to both sides.

While this is clearly good news for Google (millions of businesses and organizations bidding against each other, with all the money going to Google!), it's also great news for marketers. Even marketers who don't think of themselves as marketers.

The Kahn Law Firm probably thought of themselves as litigators, not marketers. But by using AdWords to assemble a large class of people who saw themselves as victims of a poorly labeled medication, Kahn has an advantage over other law firms. Kahn wins this round not by using their litigation skills, but by understanding the New Marketing.

Every day, hundreds of millions of people do hundreds of millions of searches on Google. Each search is its own "channel." Each search represents a distinct marketing vehicle, a chance for an individual to directly connect with a marketer.

The Sendaball Effect

Sendaball.com is a tiny company in Illinois run by two moms working from home. The idea is simple: for ten dollars, they mail an inflated, imprinted ball to a loved one. It's great for a birthday or as a get-well gag. The poor postman shows up at the door with ten pieces of junk mail and a big, bouncy ball.

That would be the end of it, and the business would fade away, except for the fact (confirmed by the company) that every single order has a 40 percent chance of leading the recipient to log on and send another order to someone else.

Four out of every ten orders leads directly to a new order from a recipient of a gift ball.

So the ball bounces.

It bounces from person to person, snaking a path through the population.

Call it an ideavirus if you like, but the sendaball effect is everywhere. It used to be far more difficult to spread the word about a song, a product, or a service. Now, the marketing is built right in. Sendaball doesn't need commercials to grow. Every time they make a sale, it turns into another sale. The product is everything the company needs to spread the idea itself.

Now it's easier than ever for organizations not to rely on external forces but to build the marketing right into the product itself.

Me-Mail, Not E-Mail

For decades, organizations built the ability to talk to the masses, to craft messages that would appeal to large numbers of people. Now, and quite suddenly, there's a huge demand for me-media, for incredibly targeted messaging that's about the consumer and what he wants, right now.

TREND 9

DIRECT COMMUNICATION AND COMMERCE BETWEEN CONSUMERS AND CONSUMERS

eBay is the beginning of a significant consumer-to-consumer connection in the marketplace. As social networks become more powerful, consumers will gravitate to each other, not just informing each other about their experiences but banding together into unions that will pressure organizations for more of what consumers want.

The New Power of eBay

Grant Stockly has built a new Altair 8800. For the nongeeks among us, it's a replica of the first home computer, the machine that came before the Apple II.

Painstakingly hand-built from hard-to-find parts, the computer recaptures an era that ended only only twenty-five years ago. Now, how to sell it?

In a different time, Grant would have to find distribution as well as promotion. He'd have to build inventory, ship it to retailers willing to carry his products, and then run promotions and ads so that people would actually visit the store to buy one.

With only ten or twenty produced a year, a product like this could never make it.

Instead, Grant built a few and put them up for sale on eBay.

He sent out a few e-mails to friends, and one of them dropped a note to the editors at boingboing.net, one of the most popular blogs in the world.

Did I mention that Grant lives in Anchorage, Alaska?

He's selling out his entire inventory at more than $1,700 each. His hobby might become a business. Either way, he's demonstrating how an interconnected world makes it easy for consumers to become producers.

Not Just a Garage Sale

While eBay has revolutionized the way individuals sell left-over Christmas presents, it has also enabled hundreds of thousands of individuals to become entrepreneurs and jobbers, selling items around the world.

Terry Gibbs is one of those people. He started collecting trains in the 1970s, and he eventually put himself through college by selling trains. eBay enabled him to quit his job and sell trains full time. He would probably never have set up a retail establishment (too much risk, too much overhead), but he now runs a successful multifaceted business out of Mesa, Arizona. Besides selling to thousands of people around the world, he's trained seven thousand individuals in the science of creating their own focused businesses.

This power extends far beyond eBay. CafePress.com allows hundreds of thousands of individuals to build customized storefronts, selling from a selection that ranges from shirts to bumper stickers. There is absolutely nothing revolutionary about CafePress. It's just a T-shirt shop. Except for the fact that more than 450,000 people are using it to run little stores. There's a store for people into boomerangs, and another selling coffee mugs that ridicule the current administration. CafePress thrives because they enable others to thrive. Their seller base already represents a population greater than that of Buffalo, New York.

Learning from Emily

Emily graduated from art school, built a MySpace page, started a blog, and began selling her art on Etsy.com, a site specializing in homemade crafts and artifacts.

I visited Etsy, trying to figure out what it was all about. Searching around, I chose to sort the paintings by "times viewed" and was completely stunned by the fact that some paintings have five hundred times as many views as others. And not because of the price or any obvious difference in quality.

Apparently the key factor is the artist. I notice that certain artists (like Emily) have hundreds of views, while others have just a handful. By my calculation, Emily has sold more than twenty thousand dollars' worth of paintings so far. (She's sold more than four hundred works of art at ten dollars or fifty dollars or more each.)

The path to create art and sell it (like just about everything else) has been fundamentally and permanently altered. It unraveled one day and was woven anew the next. Just as Josiah Wedgwood did hundreds of years ago, Emily figured out the new landscape. She built something new, something that will last. She wins. For a long time to come, Emily gets to call the shots about her art and her lifestyle and her income—because she was able to embrace the chaos in her industry and figure out how to weave a long-term asset for the future.

Need a Kidney?

Matchingdonors.com has saved more than a thousand lives in the U.S.A., all by connecting one person to another. Lori Mooney is alive today because this Web site exists.

Dealing with UNOS, the organization that runs the current U.S. method for allocating kidneys, can often lead a patient to wait a decade or more for a kidney. For Lori and other patients in some locations, that's unworkable.

Matchingdonors.com takes a different tack. Rather than relying on a centralized, controlled system for allocating kidneys (which is really the only way to handle kidneys from the recently deceased), matchingdonors.com connects people willing to give a kidney while they are alive with those in need.

A study done by the site found that 25 percent of the U.S. population was willing to donate a kidney if it would save a life. But to whom?

Matchingdonors.com is a connector. There are more than 3,575 donors already signed up, at the same time that more than 90,000 people are waiting impatiently on the UNOS list. By providing the introduction between both sides of the equation, founder Paul Dooley adds value for everyone.

TREND 10

THE SHIFTS IN SCARCITY AND ABUNDANCE

Your organization is based on exploiting scarcity. Create and sell something scarce and you can earn a profit. But when scarce things become common, and common things become scarce, you need to alter what you do all day.

What Used to Be Scarce

Things you wanted after you bought what you needed
Hard-drive space
Manual labor
Overnight shipping
Airtime
Shelf space

Long-distance phone service
Knowledge about other people

What Used to Be Abundant

Spare time
Attention
Ability to pollute without consequences
Trust
Sufficiently trained workers
Open space, clean water, and other natural resources

Companies and Organizations That Used to Be Profitable

Here are the things that used to be scarce and some examples of organizations that leveraged them:

- Things to want after you bought what you needed—companies that sold rare luxuries. Every city used to have a few fancy stores selling trinkets and baubles or the tools for various pastimes. Hermes and Tiffany's were big fish in a little pond. Now, satisfying wants is the core of our economy. The number-three bestselling item at The Sharper Image is a hands-free can opener—twenty-five dollars and it's yours.

- Hard-drive space—Seagate and EMC. The price of hard-disk space has dropped more than a hundredfold, and it's no longer a sexy business.
- Manual labor—cities filled with hard-working but not well-educated labor. Lackawanna was a good old-fashioned union town, with plenty of men willing to work hard. Now it has trouble finding employers.
- Overnight shipping—private courier services and bike messengers. FedEx turned overnight shipping into a low-cost commodity.
- Airtime—network TV. With no real competition, the networks dictated all of the terms to the big advertisers. Cable changed that.
- Shelf space—the dominant retailer in a town, and the products that commanded the most shelf space in those stores. Now, every retailer feels the effects of Wal-Mart and the Internet. The power of Macy's is vanishing.
- Long-distance phone service—AT&T, which used to charge two or three dollars a minute for a long-distance call. Today, bodegas will sell a call to Kuwait for a dime, and the Internet will do it for free.

- Knowledge about other people—Pinkerton's and demographic analysis firms, which made big bucks snooping on other people. Now, of course, it's all free on Google.

Why It Was Good to Be Bloomingdale's

Everyone has heard of Bloomingdale's, the legendary department store on the East Side of Manhattan. What you may not remember is just how vitally important this lone store used to be. If a designer was featured by the store, his career was made. The Pet Rock became a phenomenon on the strength of a successful launch in just one store—Bloomie's. Even the food shops in the basement had enough influence to change the way an entire nation thought about a fancy food item.

How could this be? And more relevant, why did it go away?

Start with location. Bloomingdale's was the only big department store in the wealthiest neighborhood in the wealthiest city in the wealthiest country in the world—at a time when where you shopped depended on where you lived. Location mattered.

Then consider advertising. When all advertising was mass advertising, it was hard for a designer or gourmet food maker or novelty inventor to afford to run ads in publications like the *New York Times*.

And shelf space: With a finite amount of space in a well-trafficked and well-publicized store, Bloomingdale's discovered that they had a great deal of power. The buyers responsible for stocking the store were able to force designers to give them exclusives and special deals. This led to a virtuous cycle: The good stuff was at Bloomingdale's because that's where the good stuff was!

Bloomingdale's success was based on the old scarcity. Scarcity of access, scarcity of shelf space, scarcity of media. When the scarcity model flipped, Bloomingdale's didn't have a chance. The foundation of their business became irrelevant, and all the marketing in the world couldn't fix it.

Why It's Good to Be iTunes

iTunes is a creature of our times—it takes advantage of the new scarcity at the same time as it embraces the atomization of the world. It rewards individuals who want a single track, not an entire album. iTunes replaces Tower Records because it increases the speed of finding what you want, eliminates the need to get into your car, and saves you precious time.

The very idea of a pastime is passé. We don't need to look for things to use up our spare time because we don't have any.

As soon as Apple put an iPod in your pocket, they had the leverage they needed to sell you music and replace an entire industry. Not because you couldn't get music anywhere else. Music is easy to find. The key assets of iTunes are Google-like selection, detailed knowledge of who you are and what you like, and the ability to save you hours of time while keeping you informed of what's new and what's now.

Tower Records served an important function when people were looking for a way to spend an evening flipping through records and had no place else to go. iTunes defeated Tower because the list of what was scarce changed so radically.

What Happened to Murder Ink?

Murder Ink was a mystery bookstore in New York City. Unlike Bloomingdale's, they didn't sell everything. Instead, they stocked and sold a tiny sliver of the 75,000 books published every year. They were in every sense a specialty store.

They leveraged a simple asset: they had books that no one else did. If you were a mystery fan and had read everything in your library and available at your local store, you made the trip to Murder Ink to stock up. Founded in 1972, the store lasted more than thirty years.

According to Publishers Lunch (quoting the *New York Times*), "The rent [for the store] has been increasing by 5 percent a year and currently runs $18,000 a month. . . . A Barnes & Noble at 82nd Street and Broadway has been chipping away at business for years [and] Amazon and eBay killed off mail-order business and sales of rare books."

That misses the point. Murder Ink went out of business because they were selling a meatball sundae. They tried to leverage one asset (the one that cost $18,000 a month—the store) instead of using the power of eBay and Amazon to leverage their real asset: ownership of the Long Tail, coupled with the ability to curate the world's largest collection of mystery books. Murder Ink refused to use the new scarcity to create an even better venture, so the world fought back.

Murder Ink could have saved time for a mystery lover in Kansas by delivering monthly picks by e-mail, with fast and easy selection of must-read books. Murder Ink could have leveraged the attention of their reader base by working with publishers to create books and events that were exclusive to the store. They could have built trust among their customers and started a regular book club. They could have empowered their intelligent, passionate workforce and turned them into bloggers and guides for the rest of the world. And they could have engaged the community and made the store a natural water cooler for an entire genre, instead of just a piece of real estate in Manhattan.

Leveraging the New Scarcity

Consider the list of items that are becoming ever scarcer and how organizations are leveraging them:

- Spare time—service firms that cook dinner or otherwise save time. Wegmans, a supermarket chain based in Rochester, New York, keeps increasing the shelf space they devote to prepared foods.
- Attention—media companies that use permission to deliver messages people want to get. Daily Candy, a simple mailing list, was put up for sale for $100 million in early 2006.
- Ability to pollute—products that reduce the impact of their creation, shipping, and consumption. LCD screens (thinner, lighter, and cooler) began replacing CRT computer monitors when the LCDs were still more expensive to purchase.

- Trust—organizations that keep their promises. JetBlue attracts business flyers, even though most employers would happily pay for executives to fly on American or United. Perhaps it is because JetBlue has the lowest rate of overbooking and bumping passengers.
- Sufficiently trained workers—individuals who don't settle and the institutions that train them. Training this generation of workers is turning into a huge industry. The University of Phoenix is part of a company that earned more than a billion dollars in profits in 2006.
- Open space, clean water, and other natural resources—Some of the poorest countries on earth are the ones that appear to have the richest natural resources. But oil revenue, apparently, is not enough to build a vibrant economy. As the vast oil reserves dwindle, countries like Qatar are rapidly investing in education as a cleaner way to generate income. Universities setting up inside the country include Carnegie Mellon University, Georgetown University, Texas A&M University, Virginia Commonwealth University, and Cornell University's Weill Medical College.

Caring About the *Other* Effects

It's hard to imagine a consumer in 1965 caring one way or the other about gas mileage or carbon emissions. Yet, as Steve Dubner points out, just one day's headlines in the *Wall Street Journal* Marketplace section (February 2007) included the following:

> *While Housing Withers, "Green" Materials Bloom*
> *Ikea to Charge for Plastic Bags ("Proceeds from the surcharge will go to an environmental-conservation group")*
> *Arctic Melting May Clear Path to Vast Deposits of Oil and Gas*
> *Emissions Caps Could Be Ruinous*
> *Biodiesel Powers Up on Financing*
> *Group Seeks Greenhouse-Gas Cuts*
> *EU Sets 20% Reduction in Emissions by 2020*

This in a newspaper that still has trouble spelling the words "greenhouse effect."

There are two reasons for this. The first is that people are more acutely aware than ever of their impact on the environment. The second and perhaps more powerful reason is that we're more aware than ever of our neighbors' impact. Which means people not only talk to each other about ideas, but they watch each other as well.

Since peer decisions are so vital in our own analysis of the world, the side effects of decisions made by others (companies, partners, suppliers, and friends) show up on our radar as well.

It's Good to Be Ideo

The number of individuals trained in human factors, mechanical, electrical, and software engineering, and in industrial and interaction design (all at once) is small indeed. And that's the skill set leveraged by Ideo.

Ideo is a design consultancy and innovation firm based in California, with offices around the world. They design products, services, and processes, ranging from low-tech toothbrushes to entire digital experiences.

By creating a small firm filled with the very best people in specific fields, Ideo leverages the scarcity of this expertise. As a result, companies like Apple, Microsoft, Pepsi, and P&G pay millions of dollars to have their products designed here.

If these companies could hire people just as good as the staff at Ideo, they would. But they can't, because talent like this is hard to find and bores easily when asked to work on a limited range of projects. Ideo can offer these employees something that a big company never could: endless variety. In the old days, Ideo would have needed patented machines or big investments in technology. Now, the only asset they need is a workforce so well trained it cannot be duplicated.

As the complexity of most industries increases, access to assembly-line workers is not the issue. Instead, organizations that attract and keep the tiniest portion of the top of the market now have the leverage to grow.

TREND 11
THE TRIUMPH OF BIG IDEAS

In a factory-based organization, little ideas are the key to success. Small improvements in efficiency or design can improve productivity and make a product just a bit more appealing. New Marketing in the noisy marketplace demands something bigger. It demands ideas that force people to sit up and take notice.

Big Ideas Can Be Simple (Like Service)

Jason Sherrill dropped me a note:

> Seth,
>
> Others I know in the IT industry tell me at least

once a month how much better a deal I could get if I bought Dell notebooks for my company rather than IBM ThinkPads.

Last night I got home at 7:00 P.M. I turned on my ThinkPad at 7:15 and heard a beep, then saw the words "Fan Error" flash on my screen, and then the computer shut down. I googled the error and determined that ThinkPads have a built-in protection mechanism to prevent the machines from running when a fan fails, so that they don't overheat and damage the processor.

My ThinkPad is two years old, but still under warranty. At 7:28 I submitted a warranty repair request through the IBM Web site. At 7:49 I received a call on my cell phone from a pleasant (but hard to understand) Indian gentleman confirming my request and to tell me that a service tech would call me within 24 hours. At 8:35 this morning, I received a call from Ken, an IBM support technician, asking me if he could stop by today between 12:00 P.M. and 1:00 P.M. to install the fan that would be arriving in his warehouse today before 10:00 a.m.

Ken showed up at 12:40 and was gone by 12:55. In less than 24 hours I went from a nonbootable system to a fully repaired notebook without ever having to pick up a phone.

Friends, that's why I continue to pay a premium for my ThinkPad notebooks.

Choosing to organize an entire enterprise around service is a big idea. There isn't room in the marketplace for every competitor to do this, and the first major provider that does establishes a bar that the others rarely reach. As a result, word spreads, and competitive insulation results.

The BlackBerry Is a Big Idea

In 2000, I spent a year working with a venture-capital firm in New York. Every Monday, the VC partners met in the all-important partners' meeting, where decisions about investments got made. This firm had more than a hundred million dollars to invest, and hundreds of business plans were flowing in.

Soon after the new conference room (complete with Internet access) was finished, we discovered that not much was getting done at the meetings. Everyone was so busy sending and answering e-mail that we were all ignoring each other. Quickly, we passed a rule: no laptops.

We were incredibly productive for about a month. Then, at one meeting, Fred seemed a little distant. He kept staring at his lap. After an hour or so, curious, I walked over and caught him. He was busy playing with his BlackBerry.

Several hundred thousand machines and billions of dollars later, the BlackBerry is now essential in many industries.

Jaded business people spend time, unprompted, comparing their devices. Upgrades are a necessity, not an option. The big idea? Connection, a basic human desire, upgraded.

Is It the Marketing?

Yes, if you describe marketing as a big idea. Yes, if creating something essential is your strategy. BlackBerry profits not because of their advertising, but because their big idea is in sync with the times.

The Drawers

The world of art has been changed by the big idea as well. From a long tradition of craftsmanship (think about the masterpieces at the Louvre), art has made a transition to a postphotography world. Now, instead of being a measure of how well you can use a paint brush, success in the world of art is how compelling your idea becomes.

Jana Napoli was in New Orleans after Hurricane Katrina, and she spent her time collecting more than seven hundred discarded drawers. The drawers represented the loss of homes, of possessions, of lives. She carefully cataloged and labeled each one and then assembled them into a piece that covers an entire wall. While skeptics can argue that her work, Floodwall, involves no craft, that it's not really "art," the fact is that it is powerful and evocative. See the piece and you will remember it for a long time.

Unlike an oil painting, the idea behind her art is almost as powerful in a photo or even in a paragraph that brings the image to the mind's eye. It is an idea that can travel, that spreads.

The End of the "Big Idea" in Advertising

There's a difference between a big idea that comes from a product or service and a big idea that comes from the world of advertising.

The secret of big-time advertising during the sixties and seventies was the "big idea." In *A Big Life in Advertising*, ad legend Mary Wells Lawrence writes, ". . . our goal was to have big, breakthrough ideas, not just to do good advertising. I wanted to create miracles." A big idea could build a brand, a career, or an entire agency.

Charlie the Tuna was a big idea. So was "Plop, plop, fizz, fizz."

Big ideas in advertising worked great when advertising was in charge. With a limited amount of spectrum and a lot of hungry consumers, the stage was set to put on a show. And the better the show, the bigger the punch line, the more profit could be made.

Today, the advertiser's big idea doesn't travel very well. Instead, the idea must be embedded into the experience of the product itself. Once again, what we used to think of as advertising or marketing is pushed deeper into the organization.

Yes, there are big ideas. They're just not advertising-based.

Market Leaders Are Threatened

If a market leader got there via the Old Marketing, he probably had advantages in distribution and manufacturing. And in the old days, those advantages could be leveraged for a long time—decades or even longer.

Today, though, distribution is more permeable and easier than ever to get around. And manufacturing is often a hindrance, not an advantage.

So Puma or Adidas can make a significant run at Nike, even though under the old rules, Nike ought to be untouchable. So Sennheiser and Shure can threaten Bose's significant market power.

A big idea can spread so far and so fast that the market leader cannot stop it.

Under CEO Bob Nardelli, Home Depot alienated customers, investors, and employees. Even though Home Depot had huge market power and a significant retail footprint, customer dissatisfaction was enough of an opening for the more agile Lowe's chain to make significant inroads. A simple idea, pursued relentlessly (women welcome here), was enough to transfer billions of dollars of market value from one company to another.

TREND 12

THE SHIFT FROM "HOW MANY" TO "WHO"

As we saw earlier, marketing is often like a funnel. Attention is shoveled in at the top and, over time, sales come out the bottom. The funneling process sorts the wheat from the chaff, separating those who can buy from those who either aren't interested or can't afford to participate.

This focus on mass is understandable if you assume that all consumers are pretty much the same or if you can't tell them apart. The thing is, they aren't, and you can. Now, for the first time, marketers can focus on who is hearing (and talking about) their message, and they no longer use mass as a placeholder.

Panning for Gold

The secret of finding gold by the side of a river is to process a lot of silt. The more silt you shake through your pan, the greater the chance of finding gold.

So marketers work hard to find as much silt as they can. They look at ratings and Web rankings. They pay extra to be on the Super Bowl or on the Yahoo! home page. No one ever got fired for buying a billboard in Times Square.

It's hard to overestimate the obsession with size that is embraced by most traditional marketers.

Combine this with a real resistance to finding a better pan. Gold prospectors are superstitious, and they like to stick with what worked in the past. If you're used to having masses of prospects in front of you, it's okay to run ads that aren't particularly efficient. Just as a store in a busy mall doesn't have to worry about converting every browser into a customer, high-traffic Web sites and advertisers get sloppy about being efficient.

One of the realities of the New Marketing is that mass is no longer achievable. Even more important: Mass is no longer desirable. Now that we can know who is coming to our Web site or store or advertising, and which ad reached them and how, we can be far more selective about what we say and why.

The Magic of AdWords

Overlooked in all the hoopla about Google is the simple power of its core revenue driver—AdWords. Those little blue boxes that show up next to each set of search results account for the bulk of the company's revenue and growth.

Do a search on "Shelby Cobra" and you'll find several ads for companies that make replicas of the Shelby Cobra.

Every day, thousands (not tens of thousands or millions, just thousands) of people do this very search. And every day, perhaps fifty or a hundred of them click on one of these links.

Compare that to the readership of *Car and Driver*, and you see the essence of the difference.

A traditional marketer buys *Car and Driver* ads because she's an optimist. She believes that if a million people (claimed circulation, counting pass-along) see her ad, then maybe, just maybe, a buried desire to own a Cobra replica will come bubbling to the surface, and they'll call her and buy a car. Maybe, just maybe, she'll sell thousands. If the ad is good enough.

The New Marketer, on the other hand, happily pays a hundred dollars (two dollars a person) for those clicks from

Google. It's a lot fewer people, of course. In fact, it's about. .005 percent as many. But the power of this medium isn't "how many," it's "who." And the "who" are people who have already demonstrated that right now, right this minute, they are focused on this car. And further, they've shown that they're willing to click on an ad to find out more about it.

But What About the People Who Aren't Looking?

The question that any marketer with an ego wants to know is "how do I reach the people who don't know about my product?" If you honestly believe that once people finally hear about your amazing product or service, they're easy converts, then your goal is to buy mass and to interrupt as many people as possible as cheaply as possible.

But you can't afford to anymore.

So you look to the New Marketing in search of cheap ways to interrupt the disinterested. But just because the medium is new doesn't mean it's cheap or efficient at doing the old kind of marketing.

Let's be really clear: The Web is the single worst medium ever devised for interrupting people who don't want to be interrupted. It costs too much, it takes too long, and it doesn't work.

Marketers who are in sync with this tool realize that it won't let you do the old thing better. What it will let you do is find people to spread the word for you. What the New Marketing enables is a process where marketers can activate the interested and turn them into campaigners for your remarkable products.

So you're going to need to give up on pushing and start working harder on cajoling.

TREND 13
THE WEALTHY ARE LIKE US

Rich people used to be all the same, just different from the rest of us. Now they're not just different from the rest of us but different from each other.

Rich people used to do similar jobs, wear similar clothes, live in similar neighborhoods, and read similar magazines. As a result, marketing to rich people was pretty easy. No longer. As the gulf between rich and poor continues to widen, and the number of people considered rich increases daily, the diversity of the rich increases as well.

It turns out that not only are the wealthy like us, they *are* us. Despite the widening gulf, there are more wealthy people than ever before. In fact, you're probably one of them. Michael Silverstein and Neil Fiske of BCG talked about this in their book *Trading Up*, and the trend has only become more pervasive.

The New Bell Curve

The old curve looked like this:

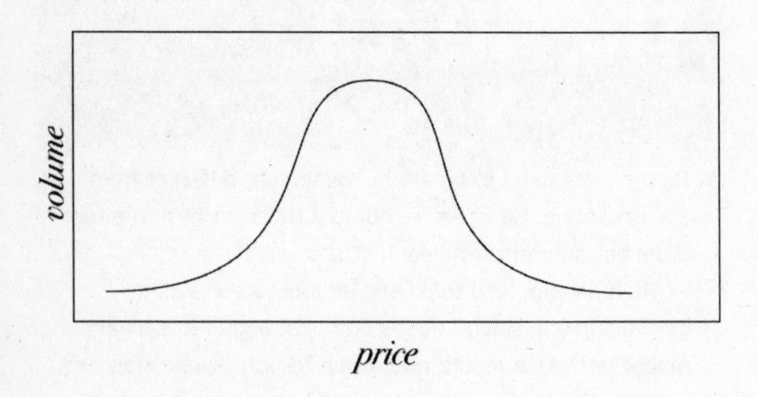

That big hump in the middle represents the typical consumer, the heart of the mass market. It represents average products at good prices. It's Sears and the corner hardware store and General Electric. Sure, there were a few people who would buy the cheapest products regardless of quality, and an equal number who happily paid a premium. The premium buyers were the folks who bought their jewelry at Harry Winston and their Christmas presents at Henri Bendel.

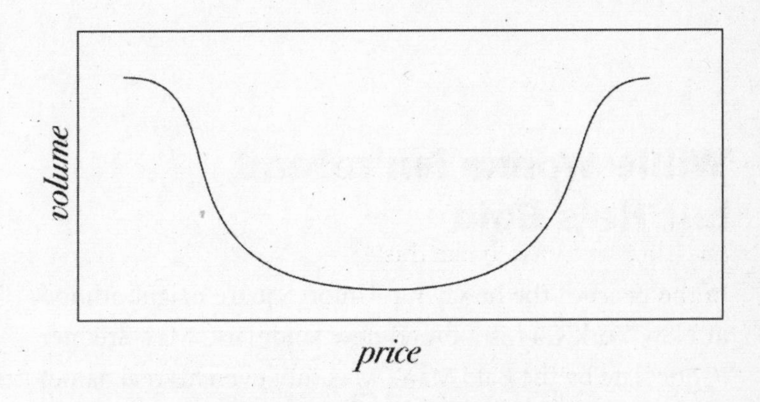

But the new reality is that the curve is reversing. Now there's a dip in the center, with humps at either end.

At the cheap end stands Wal-Mart. If I don't care enough about a product to seek out the best, then I want the cheapest.

At the other end are Viking stoves, ski lodges in Telluride, and limited-edition Puma sneakers.

Why would anyone buy something in the middle? The stuff in the middle is overpriced or under-exclusive. Given the choice, most people will avoid the mediocre middle.

Wal-Mart and Trader Joe have figured out how to sell food cheaply. The Whole Foods Market and Dean & Deluca have figured out how to make it expensive (and worth it). The folks in the middle, the Safeways and the A&Ps, are in trouble.

Willie Wonka Isn't Dead, but He's Bald

In the heart of the newly hip Union Square neighborhood in New York City is a brand-new landmark: Max Brenner (Chocolate by the Bald Man). Max (not even his real name) purportedly runs a chain of incredibly expensive chocolate cafés based in Australia. He's got almost a dozen shops there, with other outlets in Israel, Singapore, and the Philippines. The chain is profitable and growing fast.

This is the place to come if you want to order the Warm Chocolate Soup, which comes with crunch chocolate waffle balls, strawberries, and marshmallows and costs ten dollars. Or, for the ambitious, the Chocolate Mess, which is a warm chocolate cake eaten with spatulas straight from the pan, with a mountain of whipped cream, ice cream scoops, chocolate chunks, toffee cream, warm chocolate sauce, and possibly toffee bananas. It's $12.75 for one person or $37 for four.

Max's is packed, with lines of up to thirty minutes for a table. And most tables are filled with adults, not kids.

Just down the street from a Max's, you'll find the much more reasonably priced Sundaes and Cones ice cream shop, which is pretty much empty.

Why?

If I want something ordinary, then it better be cheap. I can get cheap and ordinary by the gallon at Costco. On the other hand, today's spoiled consumer is willing to pay almost anything for the exclusive, the noteworthy, and the indulgent.

Sundaes and Cones isn't cheap, and it isn't expensive. The ice cream is delicious but not revolutionary. They sell a good ice cream cone at a fair price. And that's no longer enough.

Where to Ski

Please don't tell anyone, but the skiing at Solitude is terrific. Solitude is less than thirty-five minutes from the Salt Lake City airport. They have great snow, terrific people, and absolutely no lift lines.

About an hour away, Park City has the same snow, but you'll wait an hour or more to get up some of the lifts.

Why do so many people choose to ski at the crowded ski area?

Because it's more. More expensive, more crowded, filled with more stores. It's the perfect place to feel like you're indulging yourself. It's not necessarily a rational decision, but as we've seen, most purchases aren't rational.

The typical guest at Aspen skis for two hours a day. The typical ski lodge at Telluride is occupied two or three weeks a year (and costs more than a million dollars). Apparently, few people pick a ski destination based on the skiing or because they're trying to save a dollar or two.

By being reasonable and rational and flying in the face of the democratization of wealth, Solitude suffers. If they were a lot cheaper, of course the place would be jammed. And, paradoxically, if they tripled their rates and added some amenities, it's likely that Solitude would be crowded as well.

TREND 14

NEW GATEKEEPERS, NO GATEKEEPERS

One way big organizations got bigger was by working with the other big guys. It used to be critically important to get your product into a major retailer or on the endcap of the A&P. It mattered that you were featured on a network TV show or chosen by a magazine to be highlighted. Big companies wanted to work with other big companies, and so the big got bigger.

The YouTube President

In December 2006, John Edwards announced he was running for president. Whether or not he won, the way he announced is a perfect example of the New Marketing dynamic.

Within weeks, Barack Obama and Hillary Clinton followed his lead.

John F. Kennedy was our first television president. He was elected because of the way he performed in the (first ever) televised debates. We remember Kennedy by the newsreel and TV footage of him, but that's not the way we remember Eisenhower, the president who preceded Kennedy. From a marketing perspective, Kennedy ushered in the era of political TV advertising, which means that it now costs more than a hundred million dollars to run for president. If Eisenhower spent more than $4 million, I'd be stunned.

The money spent by Reagan, Clinton, and Bush was spent for a good reason: It bought access to voters. If you wanted to reach the electorate, you had to pay a gatekeeper for access. And the gatekeeper was the media.

YouTube and other Internet media services fundamentally alter the equation. On December 27, 2006, on a slow week between Christmas and New Year's, more than fifty thousand people tuned in to Edwards' announcement on YouTube. This is far more than would have actually watched a television commercial on the same topic. (Not been *shown,* but actually *watched,* a commercial.) And Edwards was able to run a message that lasted more than half an hour—for free—with much more impact than a TV commercial could ever deliver.

It will take more than one election cycle for TV to become irrelevant. What's clear, though, is that the ability to send a message to the base—and to do it with no filters, no time constraints, and no money—is too powerful to be resisted.

The *Sick Puppies*, Free Hugs, and the Spread of Ideas

The math of record promotion has always been the same: A record label invests in promotion and recording studios and then reinvests by promoting the record heavily to radio stations.

The challenge is that there are only a few slots available for a new band on the radio. Miss those slots and nothing happens.

The Sick Puppies were a little-known rock group out of Australia. They had no major-label promotion and no radio airplay. Then, a creative soul who calls himself Juan Mann made a video. The video featured a slightly disheveled man in an outdoor shopping mall, holding a sign offering Free Hugs.

The video was posted on YouTube. The soundtrack was from the Sick Puppies. The video has been watched more than 6 million times and turned the band's records from cutouts into bestsellers.

What happened?

How did the Sick Puppies bypass the payola network of radio? Without promotional experts at their side, without distribution at Tower Records, how did they change the game? How much did it cost?

It cost nothing. The Sick Puppies benefited from a homemade video that spread. An ideavirus that spread around the world, attracting attention, not demanding it.

Now that everyone knows what the Sick Puppies sound like, the band gets another shot, a better one.

"I Know a Guy at the *New York Times*"

The old-boy network used to work. In fact, it worked really well. Write a business book and get it reviewed in the Book Review and then again in the Arts section. Contribute a piece to the Sunday business section, and perhaps even get a piece in the Sunday magazine. People who read books read the *Times,* and a blessing like that goes a very long way.

Nice work if you can get it.

Today, though, word spreads through new channels, not old ones. If you have a tech product, a glowing review from Gizmodo.com is worth more than the cover of *PC Magazine*. If you want to reach small business owners with your new accounting system, it's fine to get a paragraph or two in *BusinessWeek*, but what you really want is a hundred of your best customers to build lenses on Squidoo (a company I started) or post their experiences on their blogs.

The attraction of the old sort of gatekeeper was that the goal was straightforward, and even though getting chosen was a long shot, getting started was a simple task. Getting mentioned in the big media channel focused on your sector was a straightforward strategy to include in your business plan.

Who would have guessed that would mean having a special on the Food Network or getting a great score from the thousands of people who report to Zagats? Who could imagine that the best way to launch a multi-multibillion-dollar company was to quietly start a Web site? It's hard to remember, but Google launched without a stitch of hype. They just appeared.

Ignore the new gatekeepers at your peril.

It's easy to rejoice over the decline in influence of faded columnists and bullying editors. But that doesn't mean you're out of the woods. Sure, John Edwards can go on YouTube without anyone's permission, but twenty or a hundred bloggers and diggers are responsible for people figuring out that he's there.

The Web is the biggest haystack in the history of mankind, and you're just a tiny little needle. You might be sharp and you might be shiny, but without help, no one will ever see your Web page, listen to your podcast, or watch your video.

How to Work with the New Gatekeepers

Traditional media outlets are dependent on the PR community. Reporters may have whined about press releases and flacks, but the fact is that without them, the editors wouldn't have known what to print.

The Web is a very different place for two reasons. First, there's plenty of stuff to write about. Because everything is connected, RSS feeds send an endless stream of filtered story ideas to every blogger. Cory Doctorow, who cofounded the most popular blog (boingboing.net), gets literally thousands of story suggestions every single day. Just by reading their mail, Cory and his peers can fill an entire blog.

Second, most bloggers aren't paid for what they do. They're passionate, not punching the clock. As a result, few of them will pick up a mediocre press release or vapid idea just to fill space. You are appealing to their passion, not taking advantage of their ennui.

That means that the approaches you're used to are probably not going to work.

I got an e-mail last week that said, "We're sending out Haworth Zody chairs for review to prominent people in the Internet community and I was wondering if you'd be interested in receiving one. I know this isn't exactly related to what you typically discuss, but I thought you might be interested since nice and trendy office chairs have become a part of the business culture. . . . We only ask for two things: that you write a review of the chair and that you include a link to our Zody page. . . . You have absolute control over the content of your review whether you love it or hate it. The chair will be yours to keep or do with as you please."

The chair costs about six hundred dollars. In exchange for the chair, I'm supposed to give them a review (an honest one) and a link back to their site.

This is a great deal for them, and I'm sure that some hobbyist bloggers can't resist the free chair. But it smells wrong, and it's not a scalable, practical way to build a business. Bloggers who do this regularly will quickly lose their audience, which, after all, is the only asset they have.

If it were me, I'd take a longer-term approach. First, I'd identify the blogs that actually do have an interest in what I'm trying to have featured. And then I'd read them.

Over time, I'd start interacting with those bloggers. Submitting relevant links that have nothing to do with my company. Posting comments on the field in general. Becoming part of the circle—a contributor, not an interloper.

Then, months or even years later, when I have something relevant to add, I can send a truly personal note to someone I've interacted with. And that note will get read and thought about and perhaps even posted.

The advantage of this medium is that almost everyone is approachable, far more than someone at the *New York Times*. But approachable doesn't mean you can gain by spamming folks, no matter how generous your offer might be.

Part 3

PUTTING IT TOGETHER

The fourteen trends can seem contradictory, or too diverse to do anything with. You don't have to embrace all fourteen, but it appears that you must leverage at least a few of them. In these examples, you can see how it works in practice.

Be Like Paley

No, don't do what Paley did. That won't work. Instead, think like Paley thought.

William S. Paley built CBS from a small radio station into one of the three big TV networks, generating a fortune in the process. He understood three things:

1. People love to watch TV.
2. The FCC was regulating spectrum, and they were going to severely limit the number of stations, thus increasing the value of each one.
3. Advertisers will pay handsomely for attention, and a TV network can deliver it by the bucketful.

By investing in both stations and content, CBS built a media entity that dominated the country for more than fifty years. Paley's investments (in stations, in shows, in newscasts) paid for themselves many times over.

Obviously, this math doesn't work today. Many people would rather interact with others than watch TV. The FCC is increasingly irrelevant, and advertisers no longer are quite as patient with ads that don't work.

So what would Paley do?

He would realize that while mass media is dying, micromedia is thriving, and connecting micromedia moguls to each other can be quite profitable.

He might realize that Google has decimated the power of a traditional gatekeeper like CBS. He might instead invest in building direct, personal, permission-based relationships with millions of people—to become a new kind of gatekeeper, one that isn't supported by the artificial scarcity provided by the FCC.

Paley would certainly understand that attention must now be earned. It can no longer be demanded.

And I have no doubt that he'd learn a lesson from his peer Walter Annenberg, the founder of *TV Guide*. When Annenberg sold *TV Guide* for more than three billion dollars, it was worth far more than any of the three TV networks. Information about the content appeared to be worth more than the content itself.

Most of the existing gatekeepers (TV networks, newspapers, large retailers, business-to-business sellers with large sales forces) are going to fight the new realities. These gatekeepers are going to work hard to force the world back into the configuration where they used to thrive.

A few, though, will understand that getting in sync with the new systems is far easier and more profitable than the old ways ever were.

Thinking About WordPerfect

Years ago, Microsoft was dominant in operating systems (DOS was on almost every business desktop), but spreadsheets belonged to Lotus, and word processing was the domain of WordPerfect. In fact, WordPerfect had so much market share that it was considered untouchable. Millions of office workers had memorized the function keys necessary to run the software and were in absolutely no hurry to switch.

When Microsoft introduced Windows, WordPerfect Corporation hesitated. They saw no reason to support Microsoft's effort to sell an entirely new operating system, and they (rightly) pointed out that version 1.0 of Windows was pretty lousy. So they reinvested in their DOS version.

This is a classic example of willing the world to match your offerings, as opposed to the other way around. The folks at WordPerfect didn't want Microsoft to succeed and didn't want to support Windows for a variety of corporate reasons. And they felt that withholding their support would make it more likely that the world would stay as it was.

Microsoft took advantage of this mistake and pushed its word processor, Word for Windows. It didn't sell well for more than a year, but as the Windows OS gained traction, WordPerfect never had a chance.

The "operating system" for marketers is now fundamentally changing. It doesn't matter how big your market share is today. If your product and your marketing are optimized for the older model, you will be defeated by the relentless tide of the New Marketing and the products and services that are designed for it.

AOL's Strategy vs. a Strategy That Might Work for You

How AOL got big: They set up hundreds of promotional partnerships. They put CDs in magazines, CDs at the cash register of the bookstore, CDs at BestBuy. They ran ads in *BusinessWeek* promoting *BusinessWeek* online (which lived on AOL). AOL sent out so many CDs that one collector in California has amassed 404,000 discs.

And it worked. It worked because AOL did enough partnership deals, leveraged enough brand names, and interrupted (and annoyed) enough people that they got their idea out there.

It won't work for you.

It won't work because you don't have enough time and enough money to do enough deals to interrupt enough people to get your idea out there.

Is there a company that could possibly raise enough money and buy enough ads to do enough deals to repeat that success? While it's possible, it's not likely. Microsoft couldn't do it with the Zune. Wal-Mart couldn't do it with their MySpace knock-off. AOL couldn't do it with their Netscape brand, either.

What almost everyone else who has succeeded in the last decade has done is work the grapevine.

They've created ideas that are worth spreading and made it easy for people to spread them.

Amazon.com did deals with plenty of startups (like drugstore.com), offering them promotions in exchange for cash and stock. Alas, the promotion didn't work for drugstore.com. (It generated some cash for Amazon, but not enough to make it worth their time, in my view.)

The promotion didn't work because just knowing about something isn't the same as being excited about it or using it or, best of all, talking about it.

I've wasted more than my fair share of time trying to persuade big organizations to partner with me on various projects. Magazines, Web sites, publishers—these are organizations that have a great deal of permission with their users but are loathe to use it to create win/win solutions. And these organizations are conservative. And sometimes greedy. And they take forever.

Almost without exception, the copromotions don't work. They don't work because the offering has been so watered down by committees that what is being offered is dull at best.

On the other hand, products, services, and marketing campaigns that are forced to stand on their own tend to be more remarkable. They have to be, the creators figure, because without a helping hand from a wealthy partner, that's all they've got. It's their only chance.

Fotolog.com is a simple photo-sharing Web site. Fotolog is ranked number twenty-eight in the United States in terms of traffic among all sites. Even though it was started by Scott Heiferman, one of the most connected Internet entrepreneurs in the world, it was launched with no partnerships, no external promotion. Nothing but a good idea (a great idea) that was designed to spread.

If Scott had gone to Yahoo! or MSN or Kodak to launch Fotolog, he would have lost months and months negotiating. And he would have had to add features to make his partners happy. And the site would have failed.

Here's the lesson: "Go big or stay home" is bad advice. There are no fairy godmothers. If you want to thrive, you need to do two things:

Make something worth talking about; and

make it easy to talk about.

These attributes of success have nothing to do with budget or scale or corporate will and everything to do with a bottom-up strategy of making good stuff for the right people.

You can dream of the AOL strategy or the Oprah strategy or some other strategy that involves vast amounts of cash or vast amounts of attention. Far more realistic (and profitable) is to ignite your networks. To create a story that spreads from person to person, from blog to blog, that moves through a community and leaves an impact as it does.

What's Happening in Sidney?

Sidney, Nebraska, is like most small towns. Some people there are absolutely fed up with the local government.

In this town, though, someone has decided to do something about it. Anonymously.

Using the power of a blog, he has created a site (for free) that makes it easy for people in town to vent their feelings about the misadventures of the local government. Find it here: http: / / sidneytalk.blogspot.com. After just a few weeks, the site has received tens of thousands of visits. It's not unusual for a post to have ten or more comments, enabling conversations to take place in public for the first time ever.

This blog isn't some sort of Platonic ideal of democracy. Instead, it's more of a street fight. It's loud and fast and not particularly pretty. But it works. By the time you read this, the blog might be dead. It doesn't matter. The very nature of blogs (free and sometimes anonymous) means that they come and they go. The effect of this sort of public exchange, though, is to quickly focus the attention of those in government.

You can bet that the most devoted readers of the blog are the very people who get talked about.

The Very Best Sound

In 1989, I spent about six months shopping for a really good stereo system. I could finally afford it, I'd always wanted it, and it was time. I visited more than eight stores in Manhattan, auditioning speakers and amps and cables. I finally settled on the Celestion SL700 speaker. I found a pair of demos (with a little crease on the back) for about $2,000. The retail price at the time was just over $3,500 for a pair.

The Celestions were among the very best speakers available. They regularly showed up on most top-ten lists. This was an outrageous amount of money to spend for a speaker that was about a foot tall. And they were worth every penny. I listened to them for years with great joy.

The high-end audio market has changed dramatically since then.

The democratization of the wealthy means that there's very little room for $3,500 speakers anymore. Today, top-rated stereo speakers cost between $20,000 and $80,000 a pair. That's an increase of more than 1000 percent in less than fifteen years.

At the same time that the market has split into extremes, the retail distribution has changed as well. There are no longer dozens of stereo stores in most cities. In fact, outside of major metro areas, it's hard to find even one audio salon. Instead, a significant chunk of the market has moved online.

If you wanted to be successful in this market twenty years ago, you needed to market products of above-average quality and have them sold and supported by the largest dealer network you could acquire. Market dominance belonged to mainstream companies that could produce reliable, fairly priced products. Just about every dealer carried the same ten brands.

Today, the center of the market is rapidly hollowing out. Instead, success belongs to the exceptional products. Exceptionally expensive, exceptionally designed, or exceptionally great sounding. The word of mouth online lives on sites like audiogon.com and hearthemusic.com. There are blogs and discussion boards and articles about the most extreme items.

For $30,000, you can buy an amplifier that is so hot, you'll burn your hand. For $2,000, you can get speakers that require not much more than the power from a tiny battery to get awfully loud. You can get cables that cost more than $50 an inch.

And the items that sell are the ones that are considered the best in the world. Best in the world at sound, or best in the world at impressing your date. With every single stereo component available to every consumer anywhere, distribution no longer matters. Advertising no longer matters. Instead, the market belongs to companies offering products that are either a remarkable value (twice the quality at half the price) or remarkably weird (like the $17,000 CD player).

The marketing tools are different, and that has changed an entire industry as a result.

Inside vs. Outside, Ours vs. Everyone's

In the world of Old Marketing, sharing media makes perfect sense. No plumber creates his own Yellow Pages just because he's afraid of other plumbers having ads in the same place his run. The smart plumber understands that the Yellow Pages directory works precisely because everyone else is in it.

The same thing goes for TV ads. It would never make sense for an advertiser to have his own TV channel because no one would watch it. Sharing media is a no-brainer.

When the New Marketing arrived, a big part of it was the ability to roll your own. Each company built its own Web site, spending more and more money creating an online environment. The goal, whether you were selling clothes or records or expensive business-to-business services, was to create a destination, a place where people would come and not leave.

This has become an article of faith for many businesses, including those that don't do a lot of their business online. These organizations don't want to send people away from their site; they want to claim users for their own.

YouTube has demonstrated how ridiculous this strategy is. Either you can run a video on your site, or you can run one on ***the*** site, where it belongs to all of us. Which will get seen more?

Epinions.com makes it easy for your users to post reviews in a place where everyone will see them, and it saves you the trouble of building your own review software. Squidoo, a site I helped create, takes this a step further. Here's a place where nearly 100,000 people have built pages containing links, reviews, pictures, comments, rankings, and more. Any organization in the world can encourage their users to post pages here about products and services. It's like the Yellow Pages, except it's free.

But many organizations resist the YouTube/Epinions/Blogger/Squidoo call. They want to build their own, to control it, to do it themselves. After all, they argue, this is a strategic asset, and they shouldn't be sending users to a public place, a place they don't control, when they can try to keep them on the site instead.

Except it doesn't work.

It doesn't work for two reasons. First, you're not in charge. People don't care about you. They care about themselves and their community and their audience. Second, just like the Yellow Pages, these sites work better when they're centralized and connected, not when they are spread out and disconnected from each other.

In or out. Us or them. The New Marketing demands that you enter the public square and enable conversations, not isolate people from each other.

How the New Marketing Makes Change Happen

In the world of Old Marketing, change comes from the top. Organizations decide to sponsor the Super Bowl or put a free prize inside the cereal box because a senior manager approved the idea.

A large part of the staffing of most media companies is salespeople. You need salespeople to sell Old Marketing because without a sale, nothing happens. The cabal of ad agencies, sales reps, and retailers pushes many marketers into investing significant resources in the ads we're all used to.

Once you're on board, it's easy to become a zealot. Sure, a Valassis salesperson is responsible for a company initially running coupons in the Sunday paper, but after that first successful insertion, the idea of running plenty of coupons became part of the organization. Most of the tactics of the Old Marketing regime are deeply ingrained in large organizations today. The people in key marketing posts got those jobs because they were good at these tactics. These people understand that the best way to get promoted is to do what their boss did—which is interrupt potential consumers with ads and sales calls that get them to buy what your factory makes.

If you work for a company that's used to doing traditional marketing, then somewhere up top, someone has a big hammer. The hammer is traditional marketing spending. And every marketing problem looks like a nail.

This situation has led to a myth embraced by many companies that would like to enable the New Marketing. The myth is that you need to sell big deals to senior management to succeed. So Google and YouTube and Gather and Yahoo! all have business-development people who call on CEOs and marketing VPs in an effort to sell them on using the new platforms.

I know because I did the same thing at Yoyodyne in the late 1990s. We sold to Procter & Gamble and American Express and dozens of other huge companies. And I know because I did the same thing at Squidoo.com. We did deals with *Sports Illustrated, Rolling Stone,* the March of Dimes, and other large organizations. The thinking goes like this: If you can sell senior management on these new tools, the managers can leverage their organization and dramatically affect your platform as well as their goals.

Except it doesn't work that way.

The first problem is that in order to get companies to embrace the new tools, New Media companies often wreck their offerings in the process. They create animated banners, or they cripple the sharing functions built into their products. The very thing that made them worth working with in the first place starts to fade away.

The second problem is that, at least for now, most New Media companies don't have the market power or scale to get anything more than table scraps from the big companies.

Here's the thing: New Media channels like blogs and social networks work in a completely different way. They don't succeed because a CEO embraces them. They succeed because individuals embrace them.

There are 80 million blogs today. Google has a very successful one, and Microsoft did, too. And there are literally hundreds of thousands of companies now blogging quite successfully—but no salesperson ever showed up to make the sale.

Instead, most New Media platforms migrate to work in the laptops of people who are already passionate about the medium.

The lesson we learned at Squidoo was profound: We stopped making business development deals that required senior managers to embrace our new platform. They won't. They won't because it feels too speculative and they have trouble grasping the texture of this medium. Instead, we've discovered that "real people," the folks who care enough to develop pages on their own time, are the ones who will lead their organizations into this new world.

Television had no choice. TV ads are expensive to produce and expensive to run. So ad agencies had to hire salespeople and sell big corporations on spending millions to get on TV.

The Web isn't like that. When George Wright decided to launch the "Will It Blend?" series on YouTube, nobody told him to; no salesperson came in for some meetings and sold him on it. George invented it, they filmed it, and then his webmaster just posted a few on YouTube.

The March of Dimes runs a site called shareyourstory.org. The site collects real-life stories of parents who have had their children treated in the natal intensive care unit of a hospital. This site offers support for families who really need it. And it wasn't started by the CEO of the March of Dimes. It came from a blogging volunteer.

Getting out of a top-down mindset is a key step in embracing the New Marketing and in building a story that works. Because the media demand an authentic story, it's almost always going to come from someone in the trenches (like Robert Scoble when he worked at Microsoft), not from the top (Steve Ballmer still doesn't have a blog).

The lessons? If you're a New Media company, obsess about your users, not your partners. And if you work for a company that keeps making meatball sundaes, understand that the challenge isn't to persuade the CEO to get his act together; it's to start doing the right stuff in the trenches . . . and watch it filter up.

Case Studies

EVDO!

I had no clue what EVDO was. It's entirely possible you've never heard of it.

EVDO is the wireless technology used to allow your laptop computer to access the Internet even if you're not in a Starbucks. Offered by cell-phone companies like Sprint and Verizon, EVDO offers unlimited surfing in just about every major U.S. metro area for about sixty dollars a month. In order to make the service work, you need to buy a card that plugs into the slot in your laptop.

The cell-phone companies offer resellers huge bonuses for signing up users. A two-year contract is worth more than $1,400 to Sprint, so the stakes are pretty big.

But how do you get into the business of selling EVDO cards to businesspeople?

Here's the meatball method: Open a chain of retail stores, or, if you are really ambitious, figure out how to get a big chain like Radio Shack to add the item to their inventory. Train the Radio Shack salespeople to support one more item, but since they're busy, don't expect them to be so good at it. Advertise the stores and the cards in airplane magazines and other media that appeal to well-heeled business travelers, and hope that your prices are low enough and your stores convenient enough that you earn enough to buy more ads.

As you can probably tell by now, that's a ridiculous strategy.

There's too much clutter, and there aren't enough stores. It's too hard to teach someone about EVDO and too difficult to figure out whom to talk to and then to determine what they need, and the support is just too expensive to spread around the country. If you get into a price war, then whatever profit might have followed will evaporate.

A company called EVDOinfo.com took a very different approach. They embraced the sundae and got in sync with the rules of the New Marketing.

If you do a Google search on EVDO, odds are you will come across the largest forums on the topic.

EVDOforums.com is a free bulletin-board site packed with thousands of messages on a variety of topics having to do with EVDO. Please note: They don't rank highly in Google because they manipulated the system. No, EVDO-forums ranks highly because they ***deserve*** to. It's organic search engine optimization—they attract traffic because they've built a clearinghouse for free information.

Want to get the new Verizon card to work on your Mac? All the details are here. You will find links to free software, to tutorials, even to software that was written by members of the community for other members.

Reading all the technical detail, you may realize (I did) that you don't have the time or the talent to do this yourself. If you're willing to pay $1,400 for this sort of service, you're probably willing to pay a hundred dollars or so to have someone do it for you.

It turns out that EVDOforums.com is owned by (and thus sponsored by) EVDOinfo.com. You'll see their name on every page, along with a link to their shopping site. They've been in business since 1988 and are committed to the long haul. The forums have more than five thousand registered users (which probably means that there have been half a million viewers, given the 1-percent rule). It's not unusual to find a hundred registered experts online on a slow Sunday afternoon.

Pick up the phone and call them, and you'll be amazed to discover spectacular customer service. You'll find reps who are happy to give you their names and, more important, are willing to spend twenty minutes with you to figure out exactly what you need and want. Or you can buy the card online, using a cluttered but ultimately straightforward step-by-step system.

Less than a minute after your order is processed, you'll receive an e-mail confirmation. Another e-mail shows up after the item is shipped, which will probably be that day or the next.

Why so much emphasis on customer service?

Because of the bulletin boards.

By delighting each and every customer, the EVDO folks create an ongoing reservoir of goodwill and word of mouth. Soon after you are hooked up with your new card, two things happen. First you tell all of your business buds how cool the EVDO technology is. And then you tell them exactly where to go to buy a system of their own.

It took less time and cost less money to build this community than it would have to build a minichain of successful reseller stores in a city like Chicago. The difference is that EVDOinfo.com keeps growing, keeps spreading, and keeps getting more profitable.

Replacements Stumbles

In the Google chapter, we were introduced to Replacements, the folks who have sold hundreds of thousands of dishes, silver pieces, and doodads on eBay over the years. The core of their business remains their phone and direct-sales business, and somewhere along the line, the company forgot what made them successful in the first place.

Their eBay rankings are marred by relatively frequent complaints about slow shipping and expensive handling charges. The company is not acting as though every transaction is on the record.

Compounding the problem, the company has hired a number of people to warm the chairs in the phone room. Replacements hasn't taken advantage of outsourcing (hiring people who could afford to invest the time with every customer) or trained people to go the extra mile. Instead, there's a lot of script-reading going on. Unfortunately, spoiled customers aren't prepared to put up with that any longer. In one phone conversation I had with a Replacements operator, I was put on hold for five minutes and then transferred to a voice mailbox, even though I had a credit card in hand and was ready to order.

Unreasonable?

Well, can you imagine similar treatment in a high-margin retail outlet?

This is a common problem with successful businesses (and other organizations, for that matter). They exploit a shift in the marketplace to grab a toehold, and they grow as a result. But then, when the landscape shifts, the will to change the foundation to grow again isn't there.

If the secrets to their profits are the low overhead, the cheap North Carolina location, the high margins because of long tail inventory, and the unlimited online listings via eBay, then the cost of those profits is rushed customers, relentless public feedback, and unreasonably high service expectations.

Consumers expect that Replacements will have every dish ever made, that the company will treat them exceptionally well on the phone, and that all charges will be fair and reasonable. As long as the company can exceed that (very high) bar, nothing will stop their growth and profitability. But it's not easy. If it were, everyone would do it.

Paperback Writer

In 2006, Kevin Ryan and Brian Kehew published a book. That's not unusual. After all, three thousand books are published every week in the United States.

What's noteworthy about *Recording the Beatles* is what the authors didn't do. They didn't give the rights to a traditional publisher. They didn't fight hard for retail shelf space. They didn't buy co-op ads with big book chains, and they didn't try to get on Oprah.

Instead, Ryan and Kehew managed to sell every single copy of their book (three thousand were printed) at the very profitable price of one hundred dollars a copy. And they did it by embracing the tactics of New Marketing.

First, the book is clearly aimed at a tiny, self-identified, and easy-to-find audience. This book exhaustively examines every recording session and every record ever produced by the Beatles. It's obviously not priced or written for the dabbler. The authors invested the time to talk to the overlooked EMI engineers who worked on each record. The authors did extensive detective work as well. For example, they found a picture of a dialogue-dubbing session for the movie *Help!*, scanned it, blew it up, reversed it, and discovered that the Fab Four were actually singing the mono mix of the song.

Instead of following the meatball strategy of creating something for everyone, the authors created something for almost no one. **Almost** being the key word. They knew where to find this audience. And more important, using the Web, the audience knew where to find them: (www.recordingthebeatles.com).

By self-publishing, the authors were able to accomplish several things. First, they removed a substantial "tax" (85 percent of the cover price) that a publisher charges to handle things like retail distribution, advertising, printing risk, and staffing.

More important, self-publishing took them out of a meatball factory mindset. Instead of publishing yet another book, a book for an anonymous, unseen group of consumers who would somehow find the book they didn't know they wanted, the authors found a book for the readers they already knew about. (In fact, they were part of this very audience.)

Ryan and Kehew were already part of the discussion groups, the fan clubs, the conventions—they knew this audience, and spreading the word about the book wasn't a challenge. All the advances in person-to-person communication didn't make it more difficult to promote the book; they made it easier. Instead of searching out a gatekeeper and persuading her to get the word out, they chose to create a remarkable book, a book that the community would decide was worth talking about.

It's worth noting that I discovered the book as part of a feature in the *New York Times*, but the article wasn't written until after the hardcover edition was totally sold out. The authors now have the ability to leverage their commercial success, creating a lower-cost paperback edition that can reach people who are just as fascinated but probably unwilling to spend one hundred dollars on a book (like me).

The authors now enter a new stage. They've been embraced by the core of the community; they have created an ideavirus—the ideas in their book are spreading. Now, aided by Google, by bulletin boards, by word of mouth, and by old-style media, they can profit by selling books to people who would never have known about the book if it had been published in the traditional way.

"Wait," the skeptics point out. "This is a fine idea, but it's so small. How does this get leveraged into something powerful?" Understand this: With revenue of over $300,000 to date, *Recording the Beatles* has generated more revenue than 97 percent of all books ever published. And unlike other books, most of this revenue goes to the authors.

"Impossible to Build a New Vail"

That's what Doug Donovan said about the ski area he runs. But Echo isn't really a ski area. It's a sundae, designed for a new kind of skier and a new kind of marketing.

Traditionally, ski areas have relied on location and natural features to build a following. If they were fortunate, they attracted beautiful rich people and the land was transformed into a real-estate development filled with multimillion-dollar homes that happened to have some skiing nearby. The industry has struggled for years, because advertising hasn't been effective at filling the lifts or creating the sort of frenzy that real-estate investors thrive on. As we saw in the Solitude example in Utah, a ski area without rich people and real-estate development is a ski area that's not particularly profitable.

Echo is different. A tiny defunct ski area outside of Denver, it was sold for just $700,000 in 2003. That's less than the cost of the living room of Dan Quayle's house on the slopes of Telluride.

Jerry Pettitt set out to build a new kind of ski area, something remarkable, something that would thrive in conjunction with the New Marketing. It has no fancy lodge, no fancy lunch, no slopeside housing. No skiing, really. Instead, it's a giant skate park, a mountain filled with bowls, grinds, and rails.

The mountain's Web site is wide open, inviting the (mostly) young audience to design the future of the mountain. This is the group that decided to keep the place open under the lights to nine P.M. and that argues about rock climbing and other summer activities that ought to be offered. The mountain's MySpace page has hundreds of registered friends, and the nascent bulletin board has more than six hundred registered users.

Echo is built around a big idea. It's a place where the terrain park is the entire point of the activity, not a sidelight the way it is at a traditional area, provided for the teenager who came boarding with Dad.

The tools were available to every single ski operator in the United States. But it took an outsider to realize that committing to a completely different strategy was the only way to restart the game.

In the words of Marc Vitelli, who designed the most extreme obstacle course at Echo, "The only way to learn how to stomp tricks is by going huge and embracing the occasional yard sale."

A *yard sale* is a wipeout so extreme that your gear flies off. Exactly.

Case Study: Understanding Critical Mass—Pomme and Kelly

"Critical mass" is perhaps one of the most overused and under-understood phrases in all of marketing. Critical mass is a term from physics, as we all know. It describes what happens if you compress a radioactive substance (like plutonium) just right. Just right because you need exactly the right amount of material and just the right amount of pressure and then—boom!—an uncontrollable chain reaction takes place, one that gets bigger and bigger until all the fuel is expended. This is often known as an atomic bomb.

In marketing, a chain reaction due to critical mass can have a happier outcome. It can lead to a band ending up on top of the charts, or a piece of software becoming insanely profitable for a decade. The core idea here is being popular because you're popular. It's because something extra happens, a force that compounds over time.

Sometimes the compounding force is the efficiency a company gets when they hit scale.

And sometimes, critical mass matters because people buy what other people are buying.

When Pomme and Kelly (two teenagers from the Netherlands) videotaped their lip-synced version of Aretha Franklin's "Respect," it languished for a while on their blog. Then it got submitted to an online contest, won, and became an Internet sensation. Millions watched it on YouTube. The girls were feted on national TV and enjoyed being stars for a while. They discovered the power of critical mass. The video itself didn't change, but the market's reaction to it did. What changed was its appeal to the masses. People liked it because other people were liking it.

That's why so many people use Windows . . . because everyone else does.

The other reason critical mass pays off is volume. It

drives down prices. Digital watches are so cheap because even though the first model cost a million dollars to make, each copy of that model costs a few pennies. Without a large audience, the price would be prohibitive.

There's nothing new here. But now there's a third flavor of critical mass: You need it to enable conversations. "Look, up in the sky, it's a bird, it's a plane . . ." is one of the great images of the 1950s. The crowd looks up, focused on what everyone else is focused on.

The person pointing wasn't paid to do so. He did it because he wanted to.

The people looking aren't looking because they care about Superman. They're looking because everyone else is looking.

As our society gets more connected, it is also getting more fragmented. You can find just about anything online, billions of tiny niches. But the niches that turn a profit are the ones that attract a crowd, that establish themselves as the best in the world. Critical mass is a lot smaller than it used to be.

Fax Machines

A few fax machines are worthless. Until just about every business you worked with had one, you didn't need one.

E-mail was a bust for years until the right people had it. Web sites that depend on conversations don't succeed until they cross the very difficult road of being underused. Most never make it.

For example, one Web site tried to offer advice on local vendors. You could type in your zip code and the type of tradesman you were researching (a plumber, perhaps) and you could read reviews of what others had to say. Sort of a Zagat's for plumbers.

This is a great system—if everyone is using it. If everyone is using it, then your horrible experience with the plumber gets seen by hundreds of prospective customers. It gives more power to the consumers and makes everyone more informed.

However, if no one in your area is using the system, then there's nothing to read and no reason to post. The site is stuck because they haven't come close to reaching critical mass. And it's almost impossible for a business structured like this to even hope to reach an appropriate level of traffic.

Working in the Small

Too often, businesses set out to accomplish a vision, and even though they are unsuccessful in their early efforts, they persist, hoping that "critical mass" will solve their problems. It doesn't usually happen.

For example, if you buy $200 worth of ads on Google and sell $15 worth of stuff, why will buying $2,000 or $20,000 worth of ads get you a better result? It won't. There's no critical mass at work here. Scale offers you no value, and repetition actually decreases your results. Failing for small audiences is a loud cue that you will fail even bigger with big audiences.

Compare that failure to the benefit that goes to someone who builds a system that benefits from connections. The first few fax machines weren't particularly useful, but once there were thousands in place, each new user made the system even better. This is called Metcalfe's Law, and it applies to any organization that benefits its users by allowing them to connect with each other.

The Web makes it faster and easier to connect. It lowers the level at which critical mass is reached.

Consider the case of twitter.com. This site allows users to quickly let their followers know what they're up to and where they are. That's all it does. It's fast and simple. For months, the site simmered along with just enough traffic to make it fairly interesting to those that chose to use it. Then, in March 2006, the site was featured at the SXSW (South by Southwest) conference. Within days, traffic doubled and then doubled again. It doubled every week for months, because critical mass was reached. Each user had an incentive to bring in new users, because the more users, the better Twitter worked.

The Critical-Mass Checklist

1. Do more users benefit the other users by bringing down prices or increasing the power of communication?
2. Are we falsely relying on the masses to solve problems that are obvious when we have small numbers of customers?
3. How can we lower the number of users we'll need before the benefits of critical mass kick in?
4. Does style matter? Are we betting that people will become our customers because "everyone else is doing it?" If so, how do we realistically cross that chasm?

Case Study: 37 Signals and Highrise

37 Signals is not Microsoft. That's where this starts.

The entire company is built around a few basic premises. First, they use a rapid development process to produce quick drafts of Web-based software. Then they improve it. Over and over again. Second, they are deeply engaged with their users. Every important person in the organization is either blogging, answering e-mails, or commenting on other blogs.

This tiny company (the company's founder's e-mail address is on their homepage, and yes, he writes back) focuses on the Fortune 5,000,000—the smaller organizations that make the U.S. economy run. As of March 2006, they had more than 1 million customers for their products. The company doesn't make obscure software for tiny audiences. They make Web-based programs that organize projects, track data and files, and allow entire organizations to synchronize their sales and contact work.

At the heart of 37 Signals' success is their blog, which has more than sixty thousand subscribers. There's no way that many people would read a blog about the company, which is why the blog is about the customers and their problems, not the company and its products. On the blog, key people from the company talk about rapid software development and issues relating to design. They regularly interview outsiders and generally put on a show, a show that a small (but important) segment of the world wants to watch.

When they were planning to introduce their latest product, Highrise, they posted eight times about the software. They gave screen shots and described the trade-offs they were wrestling with. They gave their blog readers first crack at signing up for a free preview when the software was ready.

Within a day of the software launch, more than eight thousand accounts were set up. Many of them were for the free, limited edition of the software, but there were also plenty of paying customers: 150,000 contacts were entered into the contact manager by users within forty-eight hours.

Please note that software marketing is a journey, not a destination. Because the software they write works better when others use it (critical mass at work), every user has an incentive to invite others on board. And because it runs online, not off a floppy, the Web encourages that critical-mass thinking to spread.

It took BaseCamp, their biggest product, five months to reach the revenue base that Highrise hit in two days. Once a customer, you give the company permission to follow up and sell you the next product—but only because 94 percent of the users of the first product are so happy with it they recommend it to others.

In addition to permission, permeability to users is a critical part of their marketing success. After one day on the market, the company faced a huge uproar about the pricing of the product. The free version was too crippled. Instead of doing extensive research and modeling, stalling, and having meetings, the CEO just changed the rules. He did it the next day (on the blog, of course) and within minutes, comments applauding the change poured in. Not only did the comments run 84:1 in favor of the change, but many of the free users (including some that were vociferous in their complaints) signed up for the newly created (paid) level of software that was created.

After thirty days, the company announced the following results:

- Over 500,000 contacts (people and companies combined) have been added so far.
- Just about 75,000 tasks have been added. That's a lot of stuff to get done.
- 130,000 notes have been added.
- Over 7,000 cases have been created.
- About 40,000 e-mails have been forwarded into Highrise (e-mail and Highrise get along great).
- And just over 9,000 files have been uploaded so far as well.

For a consumer who lives on the Web, all of these approaches seem rational and natural. Free samples, blog posts about progress, pricing that changes on a daily basis. . . . For a shrink-wrapped software company, though, it is as scary as a unicorn in a balloon factory.

Case Study: Lessons Learned from Squidoo

Two years ago, in order to better understand the changes that were whipping through the world of marketing, I started a new company. My partners and I named it Squidoo.com. Squidoo makes Google work better by allowing anyone (including you) to build a simple page (a *lens*) about an area that you care about. You could build a lens on laptop bags or the best hotels in Paris or your blog or your consulting firm.

Let me start with the punch line: In the course of a year, we spent less than five thousand dollars on marketing. And 99 percent of that paltry sum was spent on one trade show, which was a total failure, and on a case of orange rubber squids (we have extra if you want one). Despite the tiny spend, Squidoo is now the four hundredth most popular Web site in the world. Squidoo ranks higher than the *Wall Street Journal* or *Consumer Reports* or *Forbes* or ebags.com. As of mid-2007, we have more than 165,000 lenses built by 75,000 individuals.

How'd we do it? More important (much more important), how can *you* do it?

The first lesson is the biggest one. We understood that we weren't in charge. We couldn't be, because we didn't have the money to command people to listen to us. Instead, we focused on creating an environment where other people could have a conversation, and we worked hard to offer enough value that people would choose to have the conversation in our place—and to make it about us from time to time.

The second lesson? Easy beats hard. Not only are consumers overwhelmed with choices, but they're too busy to take the time to learn irrelevant details. What's irrelevant? Whatever the person decides is irrelevant. So, knowing a demographic fact about me might be important to you, the marketer, but if it's not important to me, then hey, I'm out of here. At every step along the way, we focused on "me."

The Long Tail is the third. Given enough choice, the curve of choices will get longer and fuller. Netflix makes half their rentals on titles that aren't on their top 100 best-seller list.

Computers make it easier than ever to offer infinite variety, and now consumers are beginning to demand it. Squidoo plays right into the Long Tail—on any given day, more than fifteen thousand of our pages get traffic from around the world. We've got fifteen thousand home pages producing on any given day, and the list keeps changing.

The fourth thing we learned is how slim you can run a big organization these days. We have exactly four employees at Squidoo. We're handling millions of visitors and have hundreds of thousands of regular users. Our customer service is pretty good, and our systems don't fail. To handle an audience of this size in any other medium would take ten or twenty times as many people.

But if there's one giant lesson I'd like to share with you, it's this: We're cheating. And you can too.

We're cheating because we built something designed with the Web in mind. We organized around the ideas of Web 2.0, and as a result we're running circles around much larger, much better funded organizations. With no marketing spend at all, we reach more people than almost any other brand in the world.

It's easy to come up with a list of reasons why this would never ever work for you or your brand. Which is a completely bogus way to think about it. Instead, realize that your competition isn't going to let those reasons stand in their way. Sooner or later, you are going to play by the rules of this new game—or watch the game get won by someone else. Viacom can sue YouTube, but they can't make video sharing go away. Sony can lock up their music, but the iPod and the MP3 revolution isn't going to wait for them.

It doesn't particularly matter whether or not you sell records or do recordkeeping, whether you surf the Web or sell surfboards. It's still the same math. Consumers are in charge. They're bored. They're narcissistic. And they certainly don't have the patience for your meetings or your strategy decks or your clueless CEO.

First one in, doing it right, wins. C'mon in, the water's fine.

Case Study: The Little Notebook That Could

Do you have a Moleskine notebook?

Apparently Ernest Hemingway and Picasso did. A Moleskine is a small, blank notebook that costs between ten and twenty dollars, depending on size. No moles are actually harmed in the production of one of the books, and they are very smart looking. A brand-new, blank Moleskine is filled with possibilities. It makes you feel intelligent, powerful, and bursting with potential.

But that doesn't explain the extraordinary success of this microbrand.

A Google search turns up more than 5 million matches. (That's way more than a search for Pablo Picasso, and it's just a notebook!) Perhaps the worldwide frenzy is because of the built-in bookmark or the rubber strap used to keep the notebook closed. Last year, Modo and Modo, the Italian company that has been making the notebooks for the last ten years, sold more than 4 million of them.

The secret? They've embraced the New Marketing.

The product starts with a story, a story that's told with each and every book sold. One model is called the Van Gogh and comes in a special Van Goghian color. While Van Gogh never used one, he might have, if he'd only had the chance.

The Moleskine blog (moleskinerie.com) gets thousands of visitors, as does the Moleskine lens on Squidoo. Each blog post is illustrated with a luscious picture. There are line drawings, random doodles, and even ideas for improvements. This is an analog gadget, and there are tens of thousands of gadget freaks out there improving their notebooks and sharing the results.

The thing is: Armand Frasco, keeper of the blog, had no official relationship with Modo and Modo, the makers of the Moleskine. He didn't get paid to do it. He built a blog with thousands of pages of content and created Flickr pictures and a Squidoo lens because he wanted to. He was so successful at defining the brand that the Moleskine distributor just bought his site.

The question you shouldn't be asking is, "How do I find someone like Armand?" Instead, the question needs to be,

"How do we create a product that someone like Armand becomes obsessed with?"

The Moleskine is a product with an authentic story. It's a product that enables people to create things. It's a product that is customizable, upgradeable, and discussable. And it's a product that people want to talk about. Pull out a Moleskine in a meeting and people notice.

It's not just a notebook. Notebooks are worth fifty-nine cents. A Moleskine is a souvenir, a bargain at twenty dollars.

At this point, with just a few pages left in the book, some of you are still shaking your heads. It defies your belief. You can't imagine coming to work tomorrow trying to transform your nonprofit or your church or your services company into an organization that makes stories. You can't imagine that your hard-nosed business-to-business customer, the guy in the short-sleeved shirt who does $3 million in billings with your distributorship, is going to fall for this nonsense.

If that's you, I've failed. I'm hoping that just a few readers find themselves in your shoes, and I'm going to give it one last shot before I leave you alone.

It's not nonsense.

In 2006, Pfizer launched a prescription weight-loss drug. For dogs. (Five percent of the dogs in the United States are obese; another 20 percent are on the borderline.) Fat dogs who have failed to lose weight via other means now need to take expensive medication to lose weight.

Why are these dogs fat?

They're fat because the hard-nosed, short-sleeved business-to-business buyer you are so certain is hyper-rational is actually overfeeding his dog every single day and is now about to buy medicine so the dog will eat less.

Human beings, left to their own devices, don't act like robots or rational computers. We don't all do the same things, and we don't do things for the same reasons. Given enough choices, we'll make choices. Not always the one the spreadsheet says—just the one that feels right to us. Given an authentic story that matches our worldview, we'll believe it. And given the chance to speak up, we'll do that—loudly and often.

If you want to be in the commodity business, be my guest. If you want to sell large quantities of cheap stuff, you're welcome to it. The rest of us, though, are going to grow fast using our knowledge of human nature and the New Marketing that allows people to express that nature. You're doomed to sell a slow-growing commodity only if you want to. Any product, from a cardboard notebook to an accounting firm, can now be transformed, using the tools that are available to everyone.

Case Study: Bud TV

Bud TV, as in Budweiser TV. As in a beer company that's tired of paying for fancy limos and pretty receptionists for the TV networks and wants to own its own TV network. Online.

It sounds like a brilliant scheme. Use the nearly free bandwidth of the Net to host an entire collection of video. Sponsor it yourself. Put it online and advertise like crazy. Very Web 2.0 of them.

Problem is, it didn't work. So far, AB has reportedly spent more than $40 million building and promoting the network. Yet traffic is falling as much as 40 percent month after month. According to Quantcast, in March 2007, Bud TV had traffic equal to a site that serves as a comprehensive source for sheet rubber.

Just because you want people to come to your site and watch doesn't mean they will.

A Short Note for Bob Iger

Bob Iger is the CEO of Disney, and he's got a problem.

All of Disney's Web sites (now ranked number nine

overall in traffic) are controlled by one department. That means that the movie sites and the theme park sites and the toy sites and the TV sites are all done by people who are experts at dealing with the online world. If you run a division at Disney, you need to deal with the Web department to get your site built.

Of course, this seems quite sensible. Sensible if you view the world of marketing as somehow detached from the job of making something. If we've learned anything in the last few years (and the last hundred pages), though, it's that this just isn't true.

Disney will grow faster and become more profitable when they get their businesses in sync with the New Marketing. Once they start creating products and services that actually thrive on the trends that are changing the landscape facing marketers, Disney will return to the growth that led them to become such a dominant player.

Walt Disney (the person) built a company on the back of several key trends: the rise of the automobile, the dominance of television, and the desire of kids to go to the movies. Today, there are different trends, trends that are just as powerful, if not as obvious. To delegate these interactions to a department is to minimize them, even trivialize them. Instead of relying on a staff team of experts, Mr. Iger, you ought to hire those experts to run your core businesses and to transform them into businesses that are in sync with the interactions that are driving our markets.

Here's a checklist of what key Disney managers ought to be thinking about as they invent new businesses and improve legacy units:

- **Direct communication and commerce between producers and consumers.** After my family spends fifteen thousand dollars on a Disney vacation, do I hear from you again? If so, is it with anticipated, personal, and relevant messages? Do I get treated better or worse than someone who has never been to Disneyland before? And what happens if I go to see Monsters, Inc.? Is there a follow-up?
- **Amplification of the voice of the consumer and independent authorities.** If I'm looking for movie reviews or a response to a *20/20* story, is there a Web site that makes it easy to find? If so, do you host it? And how hard are you working at treating every single visitor to a park as if she is an influential blogger?
- **Need for authenticity as the number of sources increases.** How does the company act when no one is looking? Will the Disney brand hold up to scrutiny as you enter new markets and more people are watching?

- **Stories spread, not facts.** More than almost any other company, Disney was built on a story. What's the "story" of your next film? Not the plot, but the idea that people will share after they see it. Is it authentic, or did your marketing folks dream it up after the project was finished?

- **Extremely short attention spans due to clutter.** When is Disney going to start dominating the market for three- to five-minute films (on YouTube or e-mailed or everywhere)? When I can I audition online for the local production of High School Musical?

- **The Long Tail.** Pixar is really great at making one movie a year. But who is going to fill the void of hundreds or thousands of movies a year . . . each aimed at a very specific niche? Why not partner with NetFlix and start making Civil War movies for the ten thousand Americans who would gladly watch two a week?

- **Outsourcing.** As the cost of repetitive tasks goes down, how can various parts of the Disney experience become cheaper and cheaper? Or, even better, how does the Disney experience become so handmade and insightful that outsourcing has no effect on you?

- **Google and the dicing of everything.** When I type "Orlando Wedding" into Google, Disney World doesn't show up among the top forty matches. And yet tens of thousands of brides decide to get married there every year. How can you change the experience of getting married at one of your parks so that people choose to talk about it online?

- **Infinite channels of communication.** Does Disney intend to keep all its eggs in the TV basket? How does ABC invent a hundred or ten thousand mininetworks on YouTube? How about building and maintaining a thousand advertiser-supported blogs?
- **Direct communication and commerce between consumers and consumers.** What markets and exchanges is Disney running? Why do I need to sell my collection of Winnie the Pooh autographs on eBay? How can a fan of one of your TV shows or movies turn around and make that passion into a business?
- **The shifts in scarcity and abundance.** Why do I have to wait in line to ride Space Mountain? I'd gladly trade money for the privilege of cutting the length of my stay in a park in half.
- **The triumph of big ideas.** Outside of partnering with Pixar, Disney hasn't had many big ideas lately, has it? Who is digging deep and reinventing the various media you work in? Before Walt invented theme parks, families spent almost no time there. Before he got into TV, kids watched almost none of it. What's

the next big idea about how people will interact and entertain themselves? How often do you fail? How far from the edge are you willing to go?

- **The shift from "how many" to "who."** Why does box-office gross matter in the first weekend a movie opens? Does who goes to that movie matter more than how many people go? Are you getting better at getting the right movie in front of the right people?

- **Democratization of the wealthy.** Virtually all Disney products were created for the middle class. Now that the edges become ever more pronounced, how do you stop treating those willing to spend (a lot) more as more than a niche? Not just gold-plated Mickey Mouse dolls, but fundamentally different experiences that transcend the ordinary.

- **New gatekeepers, no gatekeepers.** As the number of places you can sell continues to skyrocket, how do you leverage (or replace) the traditional retail and media gatekeepers you've become so good at romancing? Now that a blog post is worth more than a review in *Variety*, how do you reach out to the new gatekeepers (or to all of us) before it's too late?

If you keep the "new media/New Marketing" guys in a separate room, when will all these issues get addressed?

How can you avoid the politics and corporate mind-think that will lead to uniformity if you leave the answers to just a few experts?

Walt Disney didn't figure out how to dominate three kinds of media without making some mistakes. And he didn't do it overnight. His brilliance was a firmly held conviction to some core truths and a willingness to fail over and over as he strived to get in sync with the trends that changed our world.

- Lecturers out of touch
- Limited on-line presence
- Treat new ideas for marketing with contempt
- Assignments irrelevant
- hashtags? to make a trend.
- Inserting plug ins to WP
- Negative critiscism on-line

Making info relevant & accessible

→ Clarity of communication

Marketing trends

- SM still new

LI/Blog/Twitter/FB/YT Big 5

Developing a Social Media Stra

How to engage & connect

Employee SM usage

Social Media Measurement

I want the confidence to learn, try & play — speeds, tools, conventions & platforms will continually change.

Conclusion

IT'S NOT AN ORGANIZATION, IT'S A MOVEMENT

If the New Marketing can be characterized by just one idea, it's this: Ideas that spread *through* groups of people are far more powerful than ideas delivered *at* an individual.

Social change, education, new-product launches, religious movements . . . it doesn't matter, the story is the same. Movements are at the heart of change and growth. A movement—an idea that spreads with passion through a community and leads to change—is far more powerful than any advertisement ever could be.

As you consider what to do next, you're faced with a difficult choice. It's difficult because it represents giving up something you may be quite comfortable with, and it's difficult because it requires an all-or-nothing commitment.

The opportunities for optimizing your current meatball strategy have never been better. It's easier than ever to track attention and to monetize interactions. A focused, measured effort on your part will doubtless help you sell a few more widgets or get some more service contracts.

The alternative, to some, is even more enticing. And that's to create a movement. A movement around your product or the service you sell to businesses. A movement around your opposition to a war or in favor of a zoning variance. The Internet has nothing to do with what the movement is; the Net merely makes it easier than ever for a movement take place.

More often than not, movements come from out of nowhere, from small companies or impassioned individuals. That's not necessarily because they are better qualified to do the work the New Marketing requires. In fact, in many cases they're not. The reason big organizations stumble is that they can't make the commitment. They want both strategies—they insist they can have a meatball sundae. They're wrong.

There's a significant opportunity here, perhaps the biggest of your career. The opportunity is to note the distinction between an old-style organization and a powerful movement. Either choice can work, as long as you in fact make the choice. And commit.

Acknowledgments

I'm over my head in debt to Professor Will Milberg of The New School in New York. His reading of an early draft of this manuscript completely changed my approach. If you like the book, it's his fault. If you don't, it's mine.

Wired magazine had a lot to do with many of the ideas in this book. Kevin Kelly, its founding editor, wrote *New Rules for the New Economy* in 1999. He was right then and he's right now, and that book was a huge inspiration to me. Chris Anderson, current editor of *Wired,* contributed *The Long Tail* to the lexicon, and I'm both jealous and grateful.

The other Chris Anderson, the one who runs the TED conference, has not only demonstrated a very successful new business model (one that continues to grow in status and importance) but he's also used this new medium in ways that the rest of us can learn from.

My friend Steve Dennis has had a long career in traditional bricks and mortar companies and shown real insight in his resistance to fizzy "new marketing" ideas—he understands the distinction between doing something because everyone else is doing it and doing something because it works.

37signals.com and threadless.com have been living the New Marketing revolution out loud ever since they started.

Jacqueline Novogratz is at the forefront of using both the organizational and viral nature of the New Marketing to change the world for the better. Her work at the Acumen Fund and John Wood's pioneering work at roomtoread.org are clearing a path for others to follow.

Fred Wilson was the first VC to make serious bets in what is now called Web 2.0, and he and his team continue to demonstrate an innate understanding of how the company and the marketing have to be in sync. Every time I think he's wrong, I discover it was I who was mistaken.

Lisa Gansky was perhaps the first person I ever met who understood what it meant to avoid the meatball sundae, and she continues to leave nothing but good things in her wake.

I need to thank some of the best marketing blogs out there as well, because they keep raising the bar. You can find a complete list at www.toddand.com/power150, but here are some that are worth Googling: Steve Hall at Adrants, Joseph Jaffe, Jackie Huba and Ben McConnell, Tom Peters, Mack Collier, Brian Clark, Guy Kawasai, Scott Adams (yes, that Scott Adams), John Moore, Jeff Jarvis, paidcontent.org, Kathy Sierra's back issues, Scott Ginsberg, Johnnie Moore, Tom O'Leary, Michael Arrington, and, of course, Hugh Macleod.

Megan Casey has been sharing a desk with me for two years and every single day she says or does something that inspires me. Corey Brown and Gil Hildebrandt Jr. not only make Squidoo .com tick, but they also demonstrate to the rest of the world how this whole meatball sundae thing works.

Thanks to Lisa DiMona. And thanks also to the team at Portfolio, including Adrian Zackheim, Will Weisser, Allison Sweet, Branda Maholtz, Joseph Perez, and by extension Mark Fortier.

As always, this book is dedicated to my three favorite people in the world: Helene, Alex & Mo.